THE
BALANCE OF
NATURE

THE
BALANCE OF
NATURE

Lorus J. Milne

&

Margery Milne

ILLUSTRATIONS BY OLAUS J. MURIE

ALFRED · A · KNOPF : *New York*

1960

L. C. Catalog card number: 60–13433

© 1960 Lorus and Margery Milne

THIS IS A BORZOI BOOK,

PUBLISHED BY ALFRED A. KNOPF, INC.

FIRST EDITION

TO

Charles and Ruth,

whose hearts and presidential mansion
were large enough
to let us finish our chapters
close to the Lakeland sun

NOTE

While the general reader will not need to refer to the notes at the back of the book, specialists may wish to consult them for source references or as a guide to additional information on particular subjects.

Foreword

WE WOULD LIKE to share a fresh interest—one affecting every kind of life. As the news brings reports of change in the populations of animals and in the areas where they live, the change tends to fit a pattern. Some of the causes are now evident. Clearest are those situations where animal populations respond to influences from human affairs and boomerang upon us—often to our surprise.

Of some of these changes we have told in articles published in *Canadian Nature, Scientific American,* and *The New York Times Magazine.* For permission to reprint pertinent passages, we are grateful to these publications. We are grateful also to the editorial staff of Alfred A. Knopf, Inc., for many valuable suggestions.

Most of the alterations in animal numbers represent a continuing adjustment in the balance of nature. Information about them keeps coming in. Much that we mention in our pages has occurred or been noticed only in the past few years. It represents scientific news as current as yesterday.

Seldom are these changes a simple sequence from the trivial to the complex, comparable to the story begun in 1640 by the English poet George Herbert: "For want of a nail, the shoe was lost. For want of a shoe, the horse was lost. . . ." Benjamin Franklin embroidered this tale in *Poor Richard's Almanac* for 1758: "For want of the rider, the battle was lost; for want of the battle, the kingdom was lost.

All for the lack of a horseshoe nail!" Instead, the changes in animal populations begin with an event that leads quickly into a ramifying series of alterations, many of them important to man's welfare. Seemingly, no event can occur alone. Each has its effects; these lead to new events, sometimes on a world-wide scale.

Recognizing the ramifications of change can be as intriguing as meeting the friend of a friend in a distant land and realizing all over again how interwoven are the lives around us. This recognition, as it affects animal life, shows the degree to which man is now calling the tune for organic evolution and is directing his own future. What neighbors he will have, and how and where they will live, can be guessed from the changes in our world today.

We wish to make grateful acknowledgment to the publishers of the following books for permission to quote from them: *A Cup of Sky*, by Donald Culross Peattie and Noel Peattie (Houghton, Mifflin & Co., 1950); *John of the Mountains*, by John Muir, edited by Linnie Marsh Wolfe (Houghton, Mifflin & Co., 1938); *The Living Forest*, by Jack McCormick (Harper & Brothers, 1959); *Paths Across the Earth*, by Lorus and Margery Milne (Harper & Brothers, 1958); *Round River*, by Aldo Leopold (Oxford University Press, 1953); and *Sand County Almanac*, by Aldo Leopold (Oxford University Press, 1949).

We also wish to thank *The Living Wilderness* for permission to reprint the illustrations on pages 12, 21, 38, 89, 110, 117, 124, 135, 139, 146, 163, 190, 213, 244, 271, all of which first appeared in that magazine.

L.J.M. AND M.M.

Durham, New Hampshire
June 1960

Contents

*TH*E
BALANCE OF
NATURE

A MAN *who concerns himself principally with the artificial . . . can have a particular tree cut down or an ox killed at his command, and he is ever busy spinning a web of affairs. You see him hurrying across the street with rapid strides, for hasn't the Valley railroad declared a dividend? . . . We must not put by entirely the chippy singing in the apple tree, or the white clouds, for nature declares a dividend every hour—the dew drops always pay par to the summer leaves.*

WILLIAM T. DAVIS: *Days Afield on*
Staten Island (1892)

꜠꘎꜠꘎꜠꘎꜠꘎꜠꘎꜠꘎꜠꘎꜠꘎꜠꘎꜠꘎꜠꘎꜠꘎꜠꘎꜠꘎꜠꘎꜠꘎꜠꘎

CHAPTER · 1

Nature's Web

❀

THE great observer of small animals, J. Henri Fabre, had passed his eightieth birthday when he wrote the conclusion to his tenth and final volume of *Souvenirs entomologiques.* A lifetime of careful study had not sufficed to give him all the answers he sought, and he felt able to offer only "a few truths . . . tiny pieces of the vast mosaic of things." Yet he felt sure that "others will come who, also gathering a few fragments, will assemble the whole into a picture ever growing larger but ever notched by the unknown."

From close observation comes confidence of another kind. Within limits, we get to know what to expect of living things. Our human neighbors may still have doubts when, for instance, they find us daubing spots of colored paint on the backs of flower-visiting bumblebees. Won't they sting? Surely the next bee we touch will resent interference and teach us a lesson! Why paint bees anyway?

Our explanation—that we want to be able to recognize the same bee when next we meet it—satisfies a six-year-old. Without a daub of colored paint on its back, no bumblebee can be told as an individual. Yet each has its personal habits, even to the bed it seeks at sundown in a hollyhock flower behind our home. This we can prove with our daubs of paint.

Mature minds usually are far more searching: why spend time on bees? "If we learn everything about bumblebees," we reply, "we'll also know everything about each plant and each animal inhabiting our house lot." No bumblebee lives alone. Its success or failure is bound up with untold other lives. And upon the success or failure of each bumblebee on our land depends the welfare of a great variety of other creatures.

All nature is a web, each animal and plant a separate point where the strands come together. Pull at any individual, and the whole web is affected. Physicists tell us that every time a man or a grasshopper leaps into the air, the entire earth moves in the opposite direction. Our physical world shifts an immeasurably short distance as its reaction to either a single man or a grasshopper; but it moves. And the biological world reacts to the disappearance of a single bumblebee—even if the change is immeasurably small.

By following bumblebees we hope to discover some of the forces affecting their welfare. This sport is not new. A century ago Charles Darwin tried it while exploring his environment in relation to the almost overpowering implications of his ideas on evolution. How does "survival of the fittest" apply to the insects visiting flowers? What is the environment of a bumblebee?

4

In the England of Darwin's day, many naturalists were aware that bumblebees pollinated red clover. As the principal crop raised to feed beef animals, red clover was grown year after year in field after field. Its roots grow perennially. But the luxuriance of a clover field after the first year depends upon abundant seeding.

The flowers of red clover have such deep throats that, in normal years, only a long-tongued bee can claim a rewarding sip of nectar while carrying from blossom to blossom the seed-inducing pollen. Among bees, bumblebees are almost unique in having the right length of tongue. Usually they are the chief pollinators of red clover.

Darwin knew that bumblebees habitually nest on the ground. There they build little honeypots and fill them with food to be used by their young. Fields kept planted to red clover are reliable sources of nectar, and bumblebees often nest close to them. A number of bees usually build side by side, each ready to defend the loose community from attack.

In spite of these habits, a field mouse with a sweet tooth has only to wait until all of the parent bees are away at once. And the more mice a field has, the fewer young bees reach maturity.

If field mice are enemies of bumblebees, the mouse-catching owl or cat is a friend—an ally. Every mouse removed from the vicinity of a clover field materially improves the chance of success for young bumblebees. Indirectly, the presence of a cat increases the likelihood that the red clover flowers will set seed, and that the following year's clover crop will be luxuriant—fine fodder for growing beef.

In 1859 Darwin pointed out this chain between the mousing fortunes of house cats and the well-being of red clover. Then some wag extended the story, giving a reason why the British countryside abounded in felines. According to this account, the cats were kept as company by maiden ladies, and the number of these in Britain was large because so many Englishmen were away at sea, keeping the British Navy strong. That the circle of relationships was complete came from the food habits of Englishmen in the British Navy—eating the beef raised on the clover pollinated by the bumblebees which were protected from mice by the house cats kept by the spinsters at home.

Everywhere we turn, a living fabric of food relationships links nature into a web. Sometimes the mesh seems composed of only a few strands, as in the cat-and-clover story. Often it is more detailed, participated in by a wider assortment of animals and plants whose fortunes and misfortunes interact.

Those fishermen who fish for fun enjoy their sport on the very fringe of a food web in pond or stream. They angle for game fish noted for flashing speed and strength in battle. These denizens of the water world can be expected to bite on the correct lure skillfully presented, to fight to a finish, and either to escape the hook or to be drawn within reach of a short-handled net.

Fishermen and -women spend thousands of hours and millions of dollars every year, trying to make their dreams come true. What could be more natural, in these days of planned entertainment, than to build a small lake—an artificial pond—on farm land and stock it with fish appealing to fishermen?

To imitate the real thing or improve upon it, a fishpond should contain clear water and perhaps be fringed with marsh plants among which a bullfrog or a dragonfly could feel at home. Then game fish should be introduced, to see if they will establish a breeding community.

The temptation is to stock the pond with bass alone, ignoring lesser fry. But bass of catchable size are inefficient at finding food for themselves among the worms and insects at the bottom. Instead, like most game fish, they properly are fish-eaters. Their speed and strength enable them to catch and subdue smaller fish. If only bass are stocked, they will devour one another, and only a few will grow large by cannibalism. The few may mate and lay eggs that hatch. Yet as soon as the young grow large enough for the oldsters to notice, they vanish.

To feed game fish, "forage fish" should be provided. But what kind, and how many? If a bass is a specialist, developing best on a diet of smaller fish, then the forage fishes should be specialists too—more efficient than bass in feeding on bottom worms and insects, or on the microscopic plants and larger vegetation.

In the southeastern states, where pond fishing can be enjoyed through most of the year, the favorite companion for the largemouth bass is the bluegill bream—a sunfish. The two can be stocked simultaneously, one bass to ten bluegills, with as many as fifty bass per acre of pond. Although individually the predatory bass weighs three times as much as the smaller sunfish, the forage types still represent three quarters of the weight of fish in the pond and may be far more evident from a boat or shore.

Many fishermen lump young sunfish and other forage fishes together as "minnows" and regard them as valuable only for bait. Some realize that small fishes form the principal diet of the large kinds. But size is a poor guide to the type of fish.

Some "minnows" are really minnows, and reach maturity while under three inches long. Most of these are lesser members of the family to which goldfish and carp belong. Minnows either strain the microscopic algae from the water and depend upon them for nourishment, or pick out minute crustaceans that browse on algae—crustaceans just large enough for the eyes of fish or man to see.

The algae, in turn, are living bits of green; in the surface waters only they can capture the energy of sunlight and use it in making carbohydrates, fats, and proteins. These are the foods upon which the whole animal population of a lake may depend.

This food chain of alga–minnow–game-fish seems simpler than can be believed. Even when it is extended, in that the minnows capture small crustaceans which feed on algae, the transfer of nourishment from plant to predator is remarkably direct.

The true simplicity of this web of life impressed us most during a recent visit to the string of artificial lakes manipulated by the Tennessee Valley Authority. In most of these dam-held reservoirs, the water level is raised and lowered extensively and rapidly according to the needs of man. Their shores are mostly bald, protected from erosion by great slopes of cement or the irregular, fitted stones of riprap. An enameled bathtub would appear scarcely less suitable as a home for life.

These lakes of the TVA have become a major tourist attraction—a mecca for fishermen. Rowboats dotted the reservoirs we visited, and as happy anglers returned to the docks, we saw that their creels were full. Better than ten pounds of game fish—bass-like crappies—are taken from each acre of reservoir surface without affecting the supply. Removal of the largest game fish, in fact, seems to reduce cannibalism and let smaller individuals grow.

To learn more fully how many fish can be produced by a body of water managed deliberately for the benefit of fishermen, Dr. John S. Dendy of Alabama Polytechnic Institute experimented with an artificial pond, 5.7 acres in size, in Tennessee. At regular intervals he added inorganic fertilizer to maintain a rich growth of algae in the water. From a mixed population of bass and bluegill sunfish, he found that more than 850 pounds of bass could be removed annually without depleting the supply. This was 149 pounds per acre.

Many fisheries men are interested in knowing what weight of bluegills a bass must eat. On the average, ten pounds of forage fish are consumed for each new pound of bass. Yet these ten pounds must be available without interfering with the successful reproduction of the forage fishes. They must be a surplus, a "crop" removed by the bass. And, according to the experience of Dr. H. S. Swingle in Alabama, only about a third of the total weight of a fish population may be removed without upsetting the balance between the survivors and their environment. In this way, each pound of bass must represent both ten pounds of forage fish consumed and twenty pounds more left to reproduce themselves.

No one is quite sure whether forage fishes are as efficient in their feeding as are bass on a fish diet. But if the proportion of ten pounds of microscopic plants to one pound of minnow flesh is assumed, then the 30 pounds of minnows needed in a pond to maintain each pound of bass must represent at least 300 pounds of algae taken by the forage fish in their own nutrition. Again the algae consumed must be a crop, leaving still more algae to maintain the supply. If two pounds of algae must be left for every pound eaten by minnows, then 900 pounds of algae should be present in a pond to sustain each 30 pounds of minnows, and this many pounds of forage fish should be there to support one pound of bass.

The same rule of one pound in three applies also to the fisherman who catches the bass. For every pound of bass taken, two pounds more should be left in the pond to maintain the population of game fishes. These food relationships prove that for every 10-pound bass going into a fisherman's creel, 20 pounds of bass should be kept swimming around. The 30 pounds of bass—caught and uncaught—depend upon the presence of nine hundred pounds of forage fishes and 27,000 pounds of algae. Yet all too often the fisherman becomes so intent upon his sport that he overlooks the 27,-900 pounds (almost fourteen tons) of living food—forage fish plus algae—upon which his single 10-pound trophy has depended for existence.

A saucer-shaped pond is actually a pyramid of life. The great base of the pyramid is algae. The next tier, which the plants support, is minnows. The peak of the living edifice consists of comparatively few pounds of predatory fish— the bass or crappies attractive to fishermen. Year after year the anglers can pare away the peak of the pyramid without damaging its structure. But let disaster strike the microscopic plants upon which the upper tiers of the pyramid rest, and the whole organization comes crumbling down.

The advantages in Dr. Dendy's ideally managed fishpond are evident when his yield of 149 pounds of bass per acre is compared with the ten pounds of crappies per acre bringing fishermen to the TVA reservoirs. The principal difference is in the fertilizer. Wherever a reservoir is part of a river, the mineral nutrients reaching the suspended algae come quite incidentally from upstream or in drainage waters flowing directly into the man-made lake. Fishing in the reservoirs is incidental, for the prime purposes of the dams are flood control, power production, and navigation.

Most of the fishermen on TVA impoundments are delighted with their catch. Only occasionally did we encounter an "old-timer" who looked back twenty years and recalled still better fishing from the reservoirs. Then the freshly exposed soil surrounding the new impoundments was eroding, adding the fertility of the shores and hillsides to the water and giving a special boost in nutritional value. Most dam-held lakes are richest in dissolved inorganic matter during the first few years: algae increase, forage fishes thrive, game fishes reach unnatural abundance, fishing is wonderful.

As the raw shores heal over, replacement of dissolved substances continues at a slower rate. The abundance of algae shrinks, matching the lesser concentration of natural fertilizer. With it the population of forage fish diminishes, and game fish go hungry. New generations fail to develop in previous numbers. Returning fishermen may cry out for fresh stock to be added from hatchery ponds, rather than a reduction of creel limits.

No long-term improvement can be expected from releasing more game fish in an established reservoir. The water simply lacks the capacity to nourish as much life as it did initially, or to support better fishing. At a lower productivity, a new balance is struck, with a smaller population of game fish allowing a smaller but dependable harvest.

The fish in a pond or reservoir are somewhat like the ingredients in a chemist's test tube. They can be measured and studied from all sides. Yet their lives are patterned on the same master plan as those of land creatures. The prairie and the forest have their pyramids of life. The green plants provide the base and always determine the carrying capacity among the upper tiers of the living edifice. On all of these pyramids man now presses his curious finger. From each of them he seeks to take a useful share. It can be his indefinitely, so long as he is considerate and refrains from tearing the web of life.

GRASS *feeds the ox: the ox nourishes man: man dies and goes to grass again; and so the tide of life, with everlasting repetition, in continuous circles, moves endlessly on and upward. . . .*

JOHN JAMES INGALLS
in *The Kansas Magazine* (1872)

CHAPTER · 2

A Land in Balance

MORE people appreciate green grass than are aware of the microscopic algae in a pond. "Of all plants the grasses are the most important to man. All our breadstuffs—corn, wheat, oats, rye, barley—and rice and sugar cane are grasses." On dry land they take the place of algae and support a tremendous wealth of life.

Among the grasses, some are much more palatable than others to grazing animals. Mice and ground squirrels find certain grasses particularly attractive for their seeds. If given an opportunity, any user of grasses will pick and choose, taking a variety of food within the perimeter of favorites.

Probably it is no coincidence that grasses and the great grass-eaters appeared simultaneously in the fossil record of the past. No other kind of plant is so able to withstand being trampled on. No other foliage is adapted to extending

13

itself freshly after mutilation, surviving periodic attacks by grazing animals without significant loss. Man has made no mistake in choosing grasses for his lawns as well as for cereal crops.

When the Pilgrim Fathers landed in New England, a pyramid of life based on grass had long shown remarkable stability on the prairies well beyond territory seen by white men. From northern Mexico to the environs of Lake Winnipeg, from the foothills of the Rocky Mountains to the western slopes of the Appalachians stretched great grasslands: tall-grass prairies to the east, and short-grass toward the setting sun. These were home to about thirty thousand Plains Indians, whose way of life depended almost completely upon the wandering herds of bison and of pronghorn antelopes.

Only on the plains of south central Africa has man found a like concentration of large herbivores—such a wealth of animal life based on grass. Fifty million head of bison in an average year seems to be a conservative estimate for the total number, the sum of all the thousands of separate herds. Pronghorns probably were as numerous.

On the American prairies of those days, the grama grass and buffalo grass formed the green base to the terrestrial pyramid of life. On them fed bison and pronghorns, prairie dogs and mice and grasshoppers, the equivalents of the many kinds of minnows in a pond. And preying on the land herbivores, as game fish do on forage types, were smaller numbers of wolves, coyotes, foxes, badgers, grizzly bears, hawks, owls, eagles, rattlesnakes, and thirty thousand Indians. These meat-eaters took their grass nourishment at second hand.

West of the Appalachians, where two centuries ago the bison and pronghorns crisscrossed the Great Plains, other grasses now grow in field after cultivated field. On straight paved highways we whisk today between expanses of Indian corn or Caucasian wheat. For hundreds of miles docile cattle are restrained by flimsy roadside fences that erstwhile herds of bison would have shouldered through and pronghorns would have leaped at a bound. Bison country of yesteryear is all modern wheatfields and ranch lands. Grasses cover it, often in dense stands. But many of them are Old World grasses. They support a far different pyramid of life, with man at the top.

With only thirty thousand Indians, the Great Plains reached a reasonable state of balance. The area's capital wealth, the low vegetation, varied somewhat from year to year according to the amount of rain. Yet it sustained the herds as they grazed their way or battled with predators. Captain William Clark, exploring the West in 1804 on the famed Lewis and Clark Expedition, shrewdly recognized this competition in its true role: "I observe near all large gangues of Buffalow wolves and when the Buffalow move those animals follow, and feed on those that are killed by accident or those that are too pore or fat to keep up with the gangue."

The grass on one hand and the wolves on the other gave the wandering herds of herbivores a social organization and put a premium on proper numbers. A herd could be too big, or too small, or somewhere in between. If its membership surpassed a certain size, the combined need for food exceeded the supply. At first the pace of the grazing herd might quicken, leaving infirm individuals to trail behind—food for following packs of hungry wolves. Soon, however, an oversize herd divided. Then the grazing speed of each fragment could be more leisurely.

Too small a herd could not defend itself from a concerted attack by predators. When the group consists of a dozen or fewer, neither bison nor pronghorns seem able to form a compact circle, facing outward. The young cannot cower at the center, protected by their elders. Instead, the herd will panic and stampede, allowing wolves and coyotes to pick off the immature, then the slowest adults, and finally the last of the little group.

So long as bison and antelope still roamed the Great Plains, that level treeless land resembled a closely pastured meadow pocked by the almost endless "towns" of black-tailed prairie dogs. Probably more than four hundred millions of these burrowing rodents foraged among the drought-resistant vegetation. Sometimes their dike-encircled doorways were spaced a mere four feet apart. Oftener each home lay as much as thirty feet from the nearest neighbor.

No doubt the bison and antelope were more cautious than domesticated animals when stepping through the domain of prairie dogs. Cows and sheep tend to stumble carelessly into the burrows and sprain or break a leg. Prairie dogs make no attempt to conceal these holes. Instead, the black-tails regularly cut off every plant over six inches high within as much as a hundred feet of home. They gain in this way a clear view in which to detect any approaching enemy.

A hundred feet is about as far as a black-tailed prairie dog will venture from home. Any bush or clump of grass beyond that distance then becomes a hiding place for wolf or fox, coyote or bobcat, sometimes even for a golden eagle. From this fact comes the paradox that prairie dogs depend for success upon repeated returns of big grazers such as bison—their competitors for forage. Bison knock down the distant vegetation. Without bison, it is easier for the predators to approach unseen, and prairie-dog numbers shrink in spite of abundant food.

The grasses sustain prairie dogs without limiting their numbers. Instead, the burrowers are held in check by warm-blooded enemies and by rattlesnakes, which slither into their burrows, seizing young and old alike. Badgers dig out prairie dogs. Black-footed weasels follow them into their tunnels.

16

Burrowing owls and other birds of prey take a toll. Larger carnivorous mammals lie in wait. Directly or indirectly the grasses can feed all of these forms of life.

Indirectly the grasses used to nourish and clothe and warm the Indians of the plains. These men preyed on bison for meat and hides. They built modest campfires of dry buffalo chips, as the great grasslands afforded little else that would burn well. Yet the Indians could not greatly damage either the range or the meat supply upon which they depended. They made no attempt to domesticate the bison, to settle in any one place as "home." Until the white man brought horses and firearms to America, the Indians followed the herds.

In retrospect, it is impossible not to admire the balance

of nature into which the Plains Indians fitted. They lived at peace with their environment, even if not always among themselves. Their livelihood did no significant harm to the land on which they lived. Only the great climatic changes, such as brought and banished the Ice Ages, seemed likely to affect the western ranges.

To this scene came Europeans who prided themselves on their knowledge, their wisdom, their religious certainties, their tools, their weapons. They had no idea of relationships among land, plants, and animals. They could not know that, with their superior weapons, they would almost exterminate both the Plains Indians and the bison in barely a century.

The Indians had to go. They would not adopt white man's ways and respect the boundaries of property.

The bison had to go, for much the same reason. According to the Proceedings and Debates of the Forty-third Congress, in the *Congressional Record*, Volume II: "They [the bison] eat the grass. They trample upon the plains upon which our settlers desire to herd their cattle and their sheep. . . . They destroy that pasture. They are as uncivilized as the Indians."

Particularly the bison had to go, for with bison available, the Plains Indians maintained their independence against all comers. It became a matter of starving the Indians into becoming "civilized" by destroying their food resources. The Great Plains have never recovered from the change.

Without bison or antelope or Indians, the prairie dogs might have been expected to disappear. But the stockmen moved in too quickly. They shot and trapped the predatory birds and mammals that had kept the prairie dogs in check. And their livestock trimmed the vegetation more closely than ever the bison and antelope had done.

Certainly the prairie dogs throve under the new conditions. Stockmen cursed them, particularly when domestic animals broke their legs in burrow openings and died on the open plains. Soon reliable estimates appeared: 32 prairie dogs ate as much forage as a sheep; 256 prairie dogs took the grass needed to feed a cow. With strychnine-poisoned grain and later with generators of lethal gases, the ranchers did their best to eliminate prairie dogs. Today few of their "towns" remain outside of wildlife sanctuaries and national parks.

The speed of this change can be seen in Kansas, where the burrowers occupied about two million acres in 1903. By spring of 1957, prairie dogs could be found on only 57,045 acres. And of these, a fourth were scheduled for elimination during 1957. The rest were expected to go by 1960 and damage by prairie dogs would be only a memory.

Few people have questioned whether the ideal sought by man demands too high a price. The landowner strives for something unknown in the natural world: a yield of one hundred per cent. Each farmer feels justified in protecting his crops, in begrudging any measurable loss of production through animal interference. He wants to take the place of the bison, the antelope, the prairie dogs, the grasshoppers and other insects that fed on prairie plants. Man or his domestic livestock must be the only animals to harvest the Great Plains.

Toward this end, the farmers and stockmen have eliminated the bison and most antelope in the wild. Prairie dogs have become almost as rare. The big predators—wolves, coyotes, foxes, bobcats—are practically gone from the plains. The insects remain.

Prairie dogs used to chase grasshoppers, and the larger predators also kept the edge off hunger by eating insects. Now grasshoppers and other insects thrive—unless the landowner takes the place of insect-eaters also. This is part of his cost in harvesting the total yield from plants.

Often we forget the contrast between man's one-crop

fields and nature's endless variety. One crop is the ultimate in imbalance, and must be defended constantly. We overlook, too, how many animals do both good and bad to man. The prairie dog's taste for foliage conflicts with out interests, but its liking for grasshoppers is a help. Nor is it easy to realize how small an area is home to many native animals. If man removes as weeds all except his crop plants, creatures with no special liking for the economic vegetation may eat it rather than travel to remote supplies of more palatable kinds. Other animals are driven off, upsetting still more the complex balance of undisturbed prairie.

Perhaps we tend to overlook our natural allies through unfamiliarity with modern situations where a true balance can be found. One of these was discovered recently, not on a prairie but in California's avocado orchards. So perfect is biological control in this cropland that each year less than one per cent of the total avocado acreage needs pest-control treatment. Yet insect enemies are present, unnoticed although ready to attack. An experiment tried near San Diego proved how quickly they can respond to opportunity.

For a period of eighty-four days, entomologists removed by hand every helpful parasitic and predatory animal they could find in one portion of a single tree. Within this trial period, caterpillars of one kind multiplied so rapidly that, to save the leaves, it was necessary to destroy the insects one by one. Two other avocado enemies throve until it seemed that similar action might have to be taken against them. Still other mites and scale insects increased phenomenally. But all of this change occurred only in the experimental portion of the tree. Other branches of the same avocado and other trees in the same grove presented no pest problem. Man's natural allies there continued their efficient control—at no cost to the growers.

On prairie ranch land, a balance can be reached if livestock are managed carefully. Prairie dogs cease to be a menace if a good cover of grass is present—neither too much nor too little. Too much grass is a sign of failure to utilize the land economically for cattle, although underuse of this kind can displace prairie dogs entirely. Too little grass leads to more prairie dogs, and often they are blamed unfairly for the barren ground produced by overgrazing by man's animals.

Fortunately, the ranchers are coming to understand the land they supervise. Some of them now recognize an obligation both to leave it in better condition than they found it, and to retain for the future the aesthetic values to be seen in native plants and animals. As this philosophy spreads, so too will a place for a few bison and prairie dogs. With the land once more in balance, man's own economy will also be in order.

So God created man in his own image, in the image of God created he him; male and female created he them.

And God blessed them, and God said unto them, Be fruitful, and multiply, and replenish the earth, and subdue it. . . .

The First Book of Moses, called
GENESIS, i, 27–8

✦⟨◊⟩⟨◊⟩⟨◊⟩⟨◊⟩⟨◊⟩⟨◊⟩⟨◊⟩⟨◊⟩⟨◊⟩⟨◊⟩⟨◊⟩⟨◊⟩⟨◊⟩⟨◊⟩⟨◊⟩⟨◊⟩✦

CHAPTER · 3

Besting Nature

❀

A FEW people today still live in the Stone Age. They manage with a minimum of possessions, surviving on inhospitable deserts no one else wants at present. Annually their territories shrink, their numbers dwindle. The Bushmen of South Africa and the aborigines of Australia's vast arid heartland are alike in their inability to cope with modern times.

All of these Stone Age people are nomads, following a way of life that is itself an anachronism. America had nomads too when European settlers spread westward in the eighteenth century. On the rich prairie soils, which now grow wheat for the world, the Plains Indians moved with the great herds of bison. Few primitive people have fitted so well into the pattern of nature surrounding them.

No one can guess how many centuries or millennia

the Plains Indians might have continued at peace with their land if the white man had not come. Until his arrival, only one landmark was important to these people: a herd of meat animals nearer than the horizon. Exact whereabouts on the great grasslands of the continent meant little.

Respect for land boundaries was a new idea imported by the European settlers. It led rapidly to the undoing of the nomads, and to a far larger human population nourished by the same prairies. The Plains Indians, depending upon their skill in hunting with simple weapons, needed at least a square mile per person to survive. The same square mile to-day supports more than a hundred people.

Our Plains Indians were fortunate in having so fine a soil below them, but this agricultural wealth invited encroach-ment by the more advanced immigrants. Wherever the soil is poorer, the hunting area of nomads must be correspond-ingly larger—as much as four hundred square miles per per-son in the arctic tundra. On "average soil," two square miles per person seems about right. The world has about enough land of average quality to support ten million people if all were nomads.

Group after group of early men gave up nomadism and found satisfaction in social organization. They gained by settling down, laying the basis for agriculture, for domesti-cation of animals, for civilization itself. At the same time, their attitudes toward nature changed. No longer did the gods of chance play so large a role, bringing to the favored hunter a meat animal he could kill, or guiding the footsteps of the root-gatherer to plants of edible kinds. Instead, the gods of rain and frost took on new importance. Weather determined how well the plants would grow for man and his domestic animals. Man might help himself, moreover, by building dams against floods and watering his crops through irrigation. He might benefit by co-operating with nature.

Most of our ideas on living with nature came to us from

the civilizations of the Near East and the Mediterranean areas, where the people developed one of the world's two major patterns of land use. They became raisers of annual crops of seed plants, cereal grasses in particular. They relied upon domesticated animals, principally those that graze.

Certainly these folkways did not arise suddenly. Neither a grainfield nor a herd of cattle could come into existence without a long, slow evolution. It seems probable that for a long time mankind lived at the forest edge, enjoying the sunlight but finding in the open spaces very little to eat. A surprising number of edible plants and animals thrive along the boundary of woodland, whereas the forest proper produces chiefly trees.

The eminent geographer Dr. Carl O. Sauer is convinced that agriculture arose as a by-product of deliberate wood-gathering. Men became deliberate in this task when they discovered how to produce more dead trees for fuel by girdling the bark of living ones, using their new tool, the polished stone ax. At the same time, the men of the forest edge realized that the soil exposed between the trees they killed was soft and crumbly. Until the forest spread and reoccupied the land, the soft soil would produce crops of useful plants. All a man had to do to encourage growth of valuable kinds was to make a little hole with a pointed stick, thrust a branch or a bit of root from a useful plant into the hole, and squeeze the soil around it. Nature did the rest.

Along the edge of the forest, early man used the sticks he gathered as tools. Systematically he uprooted the weeds between the trees he killed before planting a new crop of useful herbs. This small advance in technique appears to be the origin of the plowing stick, an invention of the Middle East. With it, agriculturalists could extend their farming into the lightest soils beyond the forest edge, breaking the thin sod there and preparing the ground for useful cereals. After the invention of the plowing stick, the advantage of grasslands as a place for growing food became obvious.

While soil was being prepared for crops, trees and their remains were in the way. Dead, girdled trunks did not cast enough shade to interfere with agriculture, but the root-filled ground endangered man's new tool. Among the stumps of a former forest, the soil was full of obstructions on which plowing sticks caught and were broken. Why not eliminate the trees, roots and all, with fire? To burn wood except as fuel represented a new attitude toward nature.

To till the land with a plowing stick was hard work. It progressed much faster when a man had help. No doubt his wife was called upon. With a rope of plant fibers or braided rawhide she would pull the plow while her man wrestled the stick itself through the root-laced soil. The stronger his wife was, the better he could prepare the ground for useful plants. What could be more obvious to him than a correlation between the help of a wife in the years of her fertility and the improved fertility of the plowed soil? This was the sort of magic upon which a primitive man could fasten his hopes. It could become codified into tribal lore.

In some parts of India the use of women of child-bearing age as draft animals in pulling crude plows has continued to the present day. Elsewhere the women were replaced as draft animals when domesticated buffalo, cattle, and horses came into use. In India, however, the folklore survived, according to the view of Dr. Edavaleth K. Janaki Ammal, Director of the Central Botanical Laboratories at Lucknow. She traces a transfer of the superstitious connection between the fertility of woman and the fertility of the soil. When the buffalo and ox took the place of women in pulling plows, they too became sacred—a part of the magic needed to get food from the soil.

Dr. Janaki Ammal believes that most Indians who regard cattle as sacred in India today know no basis for their convictions. Nor do they realize that their faith is responsible for

25

the existence on the subcontinent of the largest concentration of cattle in the world, perhaps two hundred million animals, consuming about three times as much food as India's human population. Only about a third of these animals are used, as draft animals or as a source of dairy products. The rest wander freely, protected by religious conviction.

No such sentiment guarded trees. Two thirds of the world's people still use wood for cooking. For building materials and other needs, wood has been in great demand for many centuries. Nearly five thousand years ago the Egyptians of the Third Dynasty were drawing upon the forests of Asia Minor for timbers, using them to supplement building stone. Solomon sent "fourscore thousand hewers" to cut giant cedars from the hills of Lebanon, for construction and ornamentation of his palace and temple. Today the only remnant of these famous forests is a small protected grove at Las Cedres, on the otherwise bald slopes of the Lebanon range.

Around Solomon's time, man began to use wood as fuel in the smelting of iron. Soon the price of iron fell to a point where the metal could be used economically for plowshares, a mark of the Iron Age. An iron plowshare is so superior to a plowing stick that, when pulled by a brawny draft animal, it cuts through the matted grass roots of really heavy soils—lands on which large crops of grain can be raised. These changes—trees into smelter fuel, and iron plowshares instead of plowing sticks—came between 1100 and about 560 B.C. They led to a rapid spread of agriculture into the plains, and to progressive reduction of forest resources.

From our modern vantage point it seems strange to regard the iron industry as having made the greatest single demand on wood as fuel. Yet until the 1700's, when Europe's forests were largely depleted and the need for a substitute fuel had grown acute, almost no coal was used in iron-smelting. In his 1786 book entitled *Silva: A Discourse*

of Forest Trees, the English Royalist John Evelyn philosophized about the wood requirements of the iron industry. He noted that, although smelters usually were located close to ore pits, the land around most of them had originally been forested. "Nature has thought fit to produce this wasting ore," he wrote, "more plentifully in wood-land than in any other ground, and to enrich our forests to their own destruction."

With iron plowshares preparing heavy soils for cereal crops, grain production rose spectacularly. With it came a great increase in human population. In 1650 the world supported about 545 millions of mankind. Since 12,000 B.C. the human species had doubled its numbers five times—10 millions to 20, 20 to 40, 40 to 80, 80 to 160, 160 to 320—and was well along toward doubling them again. On the average, each doubling required some twenty-two hundred years.

The productivity of each agricultural worker became sufficient to support three people remote from the land. These others lived in cities, served there by grain ships or by wagons bearing food from adjacent fields. Time could be spent on invention and the arts. The Industrial Revolution became possible. Simultaneously, a major segment of the population lost personal contact with food-producing land. Interest in soil waned, and feelings of dependence upon nature largely disappeared.

Nature, in fact, had been harnessed. Following the tillage techniques of the Near East, men plowed the fields once a year, minimizing weed growth and providing a fine texture in surface soil, making it suitable as a bed for the germination of small seeds. Traditionally, the sower broadcast his grain by hand, depending upon the tilled soil to be hospitable to it. He measured his success in terms of how many bushels he could harvest for each bushel of grain used as seed. Yet until the crop ripened, he had little need to work in the field. Weather—one face of nature—was seen as determining the crop. Weather could mean rain in reasonable amounts at respectable intervals. Or floods and erosion. Or drought, and dust, and despair.

As fields of grain ripened, some personal intervention might be necessary to frighten away seed-eating animals such as birds. Finally the crop was ready, and people came from town to help reap the cereal heads with knives and to thresh out the edible kernels. It was hard work, but sociable. Seldom did anyone foresee ruination of the land. Usually the season was filled with happiness. For many, the annual harvest became a reminder of how well man had bested nature.

While agriculture spread its fields, domestication of buffalo, cattle, sheep, goats, and horses apparently took place. Initially these feats of husbandry seem to have been a means of getting a supply of milk, rather than of gaining meat or draft animals. Later these other uses were adopted in some areas, and greatly increased the value of livestock. Yet large herds devastate the soil.

Grazing animals appeal to man's sense of economic use for land, and support city dwellers at a minimum cost of manpower. The grazers nourish themselves on areas unsuited to cereal crops, eating weeds from soil too rocky for any plow. Only a few herdsmen need stay with the animals, to protect them from predators living in the tall grass or among the forested hills. Half a dozen boys, or even girls, can manage a large herd, leaving the rest of the human population to engage in the many activities of enlarging communities.

The ease with which grazing animals could be maintained increased without effort from the herdsmen. As wood was used for fuel in keeping man warm, for cooking his food, for metal-working, and for construction of all manner of devices (including the ships with which city folk could trade manufactured goods for food grown in more distant, less advanced communities), the forests receded and with them the predators that found shelter among the trees. So many of man's civilized activities promoted deforestation of the land that great areas of new pasture were turned over to the herds.

When trees close to large communities grew scarce, it became necessary to turn from wood to stone for buildings and bridges. Wood for fuel was imported from remote regions. At the same time, local rivers tended to disappear or be dangerously polluted. Stone aqueducts became essential to bring water from distant hills on which forests still served as traps for rain.

History books often emphasize the spread of great cities, the skillful masonry in buildings now tumbling down, the engineering marvels of aqueducts and sewers. Seldom do these volumes point out how many of the features described arose through necessity. Because of change—especially erosion—in surrounding land, communities had either to install these public works or move elsewhere.

In reading histories, it is easy to overlook the shifts in emphasis during intervening centuries. Sewers of antiquity, for example, had no relation to disposal of wastes from homes and industry. This modern use arose in Europe barely more than a century ago. Instead, the sewers of ancient cities were constructed to carry away water from ordinary storms before the rain could flood homes or undermine foundations. The land on which cities were built could not absorb the rain that fell. Tiled roofs and cobbled streets hurried the fallen rain toward the sea. Nor did surrounding hills wear a protective sponge of vegetation. From them too the water sluiced down upon the cities in the valleys and along the coasts.

The unwritten history of soil depletion, of erosion, of increased difficulty in importing enough food, water, and fuel to keep a city operating is far more difficult to establish than the names of generals and the dates of battles. Yet each great civilization of the past seems to have crumbled in turn when these problems in logistics became insuperable.

Now it is possible to read between the lines of ancient records. The soil conservationist Dr. Yaaquov Orev, in comparing present-day Texas with land familiar to him around Tel Aviv, Israel, can find much special meaning in the chapters of Genesis. He sees the patriarch Abram as a stockman of 1500 B.C., taking his flocks and his family from the irrigated valley lands of the Tigris and the Euphrates, driving them westward in search of greener pastures. Among the foothills of Canaan, to the east of the Mediterranean, Abram found what he wanted: grass, in a land not yet thickly populated. He settled there, but soon encountered a drought (Chapter 12) and then conflict over grass to feed both his own flocks and those of his nephew Lot (Chapter 13).

In Texas the stockmen were willing to fight the Mexicans for possession of rich grasslands on which to pasture cattle. Yet almost immediately the Texan prairies were as overgrazed as those in Canaan. In both areas the grass proved unable to compete with woody plants avoided by the cattle, and these shrubs soon crowded out both grass and cattle. Within a single human generation the Texan prairies became a thorn forest few would think worth battling over.

Dr. Orev finds an indication of an equivalent change right in the middle of the story about the planned sacrifice of Isaac (Chapter 22). The patriarch must have known the adjacent mountain as a grass-covered area, for he took "the wood of the burnt offering and laid it upon Isaac, his son" as they went up the slope sorrowfully together. But before the story ends, the mountain has a scrubby cover so dense that the distressed father can lift up his eyes "and behold behind him a ram caught in a thicket by his horns."

About three hundred years later, stock-raising ceased in Canaan. Jacob and his tribe sought refuge in Egypt, where food could be grown in the Nile Valley. They moved, even though doing so meant going into slavery—a status continued for nearly four centuries. Then their descendants returned to Canaan, expecting to find it again a land "overflowing with milk and honey." Instead, it proved to be densely populated, the original grasslands a tangle of scrub oaks and pistacias in which the local people collected firewood and fiber. The brush concealed wild beasts, such as the lion Samson met (Judges, 14), and grew thick enough to hide David during his days as an outlaw (I Samuel, 20). A mounted man, riding through it, could catch his long hair on low branches, as Absalom is recorded as doing (II Samuel, 18). These thorn scrubs are still the conspicuous vegetation in a land whose grassy plains once had satisfied Abram the stockman so completely.

If the Israelites were starved into slavery in Egypt, what of the principal centers of civilization between which Canaan lay? Military conquests ended the empires of Assyria and Babylonia and Egypt and Greece, but only after the land upon which the people depended for food became too unproductive to support them properly. Did Rome and the empire it controlled fall because of Alaric's guerrilla bands of Visigoths or because the grainfields of Libya turned to desert and the colonies found so much trouble raising food for themselves that they could no longer support Caesar's successors as well? An armed raider is easier to remember than a farmer abandoning his eroded fields, or a herdsman driving his flocks away from a mountain slope that has become too thorny for use.

It is only fair, however, also to praise the farmer and the herdsman. They made civilization possible. They established a pattern of land use that required comparatively few man-hours in raising enough food for large populations. When aided by the products of the Industrial Revolution, their efficiency rose to the point where each agricultural worker could support not merely three, but twenty people remote from the fields. On this base the human population rose from the 545 millions in the world of 1650 to 728 millions in 1750, and 1,171 millions by 1850. In only two centuries—not twenty-two centuries—the earth's total of mankind doubled again.

The herdsmen and agriculturalists pushed back the forests and exposed the land to a sequence of changes that end in desert unless the most conscientious and knowledgeable care is given to the soil. Until very recently the knowledge needed was not available. And today the pressure for more food for more people is often so great that knowledge is set aside. Nor can we follow the precedent of earlier civilizations. We cannot shift the center of culture to some unexploited place on earth and start afresh. Almost every area suitable for food production on this planet is already yielding bumper crops. What opportunities await our descendants on Mars or Venus remains unknown. But until space travel becomes a commonplace, we might do well to benefit from man's earlier experience by husbanding resources already within our reach.

Agriculturalists have done wonders in increasing the efficiency with which food can be raised if the future welfare of the soil is ignored. But they have not kept up with the rate at which modern medicine has saved human lives, extending the normal span of existence and increasing the number of mouths to be fed. By 1900 the world's population reached the 1,608 million mark, and in 1950 the United Nations census recorded 2,400 millions. By the year 2000 a total of 6,280 million persons is expected—two doublings of mankind in a single century! At this rate of increase, another 600 years will leave each human being barely more than a square yard of land to live on.

Economists are alarmed over man's success. Where will it lead? Can the world support all of these people and other living things as well? Already on every continent the food webs are tearing apart, often in ways that defy prediction and challenge repair or replacement.

. . . my greatest skill has been to want but little. . . . I am convinced, both by faith and experience, that to maintain one's self on this earth is not a hardship but a pastime, if we live simply and wisely; as the pursuits of the simpler nations are still the sports of the more artificial. It is not necessary that a man should earn his living by the sweat of his brow, unless he sweats easier than I do.

HENRY DAVID THOREAU: *Walden* (1854)

CHAPTER · 4

Living with Nature

For twenty-six consecutive months at the edge of the wilderness a white man in New England demonstrated to himself and the world the advantages of keeping possessions to a minimum. From July 1845 until September 1847, Henry David Thoreau lived in a shack at Walden Pond, on another man's land. He chose to subsist on what he could raise or catch by his own efforts.

Thoreau hoed beans in a clearing. He fished and thought while floating in a crude boat on the calm surface of the woods-encircled pond. But he supported no family, paid neither rent nor taxes, left no heirs, and produced great literature that could be appreciated only in the civilization upon which he had temporarily turned his back.

In the Tropics many families subsist as simply as Thoreau did, continuously following a major pattern of land use. They depend upon perennial garden plots managed by the husband and wife with the help of their children. The traditions they follow probably represent a type of tillage older than grain-raising or the domestication of animals. We have encountered remnants of this subsistence agriculture wherever we have explored off the beaten track in the mountains of tropical America and Mexico, regions in which the year falls naturally into two seasons. One is Dry Season, when evaporation exceeds rainfall. The other is Wet Season, when the converse is true.

After a few weeks of Dry Season, dead wood exposed to air burns easily. It is then that we have watched the strange birth of subsistence gardens, each rising like the legendary phoenix from the ashes of a jungle fire. The peasant farmer prepares for this event by deliberately girdling or cutting the forest trees. As soon as the brush is dry, he ignites it, and ashes fall to the soil as a quick fertilizer. Promptly the cleared space is stocked by hand with a wide variety of useful plants, often in tiers of vegetation. Vines soon creep over the ground. Beans climb the dead trees or even the cornstalks. The profusion of leaves suggests chaos, for few of the plants are in rows. Yet the soil is covered with valuable plants and may remain so for several years.

Agriculture by these primitive methods fits well into the life of the family. Some products are ready almost every day, and no sudden need for extra help arises. Whenever a bit of soil is exposed by removal of a plant, it is used immediately as space in which to start another. Often the same plant yields a variety of different products as it ages: pot herbs from young foliage; decorative flowers; fruits; starch-filled rootstocks; perhaps fibers from the mature stem, or a fermentable juice. Other plants from this culture show outstanding productivity and a long succession of products: bananas, yams, and cassava (manioc—source of tapioca starch). In consequence, the idea of a harvesttime is foreign to this style of agriculture.

We are still astonished at the number of concepts we took with us to the Tropics on our early visits, only to be hindered by them in understanding the working way of life close to the equator. Other commitments for the year limited our initial trip to the summer months of northern latitudes. We ran into Wet Season and, like the novices we were, expected that periods of blue sky and sunshine would be rare. Instead, the air was sparkling clear, washed and haze-free, the sun bright and hot. Colors were brilliant, the vegetation was lush, the animals were unobtrusive but in great variety.

On a later visit we chose to go in northern winter, from December to April, when the roads would be dry and reliably passable. But in the Dry Season of some countries it still rained, often enough to turn clay-surfaced thoroughfares into traps for anything less than a bulldozer. Between these rains, the air grew blue with the smoke from burning garden patches. Vegetation languished, and many kinds of animals reduced their activities as much as possible, as they awaited the return of Wet Season.

We were aghast to see fires lit so widely, to find valuable timber going up in smoke. Agriculturalists experienced in these lands assured us that a new network of roots would form within weeks and keep the steep soils from eroding. They insisted that methods suited to temperate zones did not work as well as the burn-and-plant system traditional in tropical America.

At the time we did not realize that on a great many subsistence farmlets, man really lives at peace with his land. He asks little from it, and does almost nothing to encourage higher yields. When the forest reinvades, he lets it come as part of a casual rotation of crops. Instead of weeding out saplings to protect his garden, he allows them to grow.

Then, toward the end of Wet Season, he girdles trees in a fresh area and starts a new garden patch. By the time his surviving children are grown and some of them inherit the land, the earlier areas will be jungle again and ready for another round of use.

A majority of these patchy gardens, whether viewed from a plane or from the ground, seem glued to mountain slopes as though applied by a paper-hanger. On the farmlet the peasant follows parental precept. He may be illiterate, and completely unaware that his system achieves a good rotational cropping of the land. Only where greater productivity is required of the land has it been necessary to codify farming procedure, to argue the relative merits of a two-field alternation between row crops and hay, as against a three-field system demanding less of the soil.

In Latin America the burn-and-plant method has been successful for centuries in all areas so remote from a center of civilization that no pressure for greater productivity is felt. Only when the peasant raises more food than is needed for his own family's subsistence is he likely to harm the land. Two imported tools are then the means for extending the life of a garden patch after its fertility has begun to slip. Why not use the steel machete to remove saplings of invading forest trees and get a few more months of use from the irregular clearing? The forest need not return at all. What food does it give? Why not use the steel ax to extend the cultivated land in every direction?

Within ill-defined limits, land tolerates alterations in the plants and animals living upon it. The population of each kind of creature rises or falls according to the demands of neighbors, including man. Rarely does a change come suddenly. Instead, the slow alteration is a "succession."

Successions long antedated man. Wind and the birds set the stage for one in every forested area by bringing in seeds and dropping them among the trees. Many of these dormant

seeds can survive a fire that ravages the forest, leaving only blackened trunks. Other seeds arrive soon after the embers cool. Brambles and saplings of fast-growing pioneer trees spring up. These are plants that cannot grow in forest shade. Yet within a few years they heal over the landscape, covering it with a delicate film of green. Among them appear a few slow-growing trees, showing that the original type of forest is reasserting itself. Very gradually these more cumbrous kinds rise up and shade out the pioneers. Once more the succession has reached its climax.

Whether in Tropical or temperate latitudes, forest and grassland are less different than might be supposed. The boundary between them is particularly vulnerable to change. In general, trees grow on land receiving enough moisture to support growth through all months when the temperature is above the freezing point. Grass, by contrast, matches slightly drier climates, taking advantage of its ability to lie dormant during regular dry seasons—usually at the end of summer. But climate is not constant, and neither is rainfall. If several wet years follow one another, the seedlings of forest trees spring up well out into adjacent grassland. They may gain a firm enough roothold to grow rapidly, shading the grass and competing so successfully for soil moisture that the grass dies. Or a sequence of dry years may kill the boundary trees, letting grass spread between the leafless trunks.

Long ago the Plains Indians of America observed this uncertain boundary between forest and prairie. They knew too that only the grasslands were hunting grounds, the home of bison. Indian fortunes could be improved by extension of grasslands. To help themselves, the Indians deliberately set fire to the dry grasses along the forest edge. Repeated scorching often killed back the woodlands, enlarging the prairies and the potential supply of meat. Today no one knows how much of the long-grass prairie stretching east of the Mississippi was in trees before the Indians began changing the landscape.

When viewed in modern perspective, most of man's activities are seen to affect plants in this same direction—inducing successions toward vegetation tolerating more and more arid conditions. Forest becomes prairie. Prairie becomes chaparral, or thorn scrub, or outright desert. Indeed, the recent invention of reforesting land is almost unique in pushing plant successions the other way—toward types requiring more moisture.

In keeping with the more uniformly high temperatures of the Tropics, succession is faster in a jungle clearing. If trees are permitted to return, a new generation is soon thrusting its roots deeper into the soil, reaching nutrients far below levels penetrated by man's vegetables. In burn-and-plant methods, the trees themselves serve during fallow years to haul up the dissolved minerals and contribute to soil fertility.

Curiously enough, forested land everywhere is expected

to maintain itself without human help. Few have suggested that fertilizers should be added to jungle or woodland soil to improve the crop of trees. Yet trees provide by far the greatest harvest man takes from the plant kingdom. This is made possible without use of fertilizers because of the un-counted minute creatures inhabiting the moist, loose soil of the forest. Each nourishes itself on decay products while making valuable substances available to the roots of trees.

Grasslands, from which man takes a smaller yield of vegetation, soon become depleted unless the nutrients are replaced. Moisture there is too scarce for soil bacteria, fungi, and animals to be continuously active in decay opera-tions. The mineral particles, moreover, tend to pack to-gether, limiting the penetration of oxygen needed by the many forms of life.

The living population of agricultural soil reaches aston-ishing dimensions: 567 million bacterial cells, 28 million fungus plants, 28 million single-celled animals, and nearly 23 million algae to the ounce. An acre of ground, to the depth ordinarily turned in plowing, has from five to twenty-five tons of this invisible life so important in maintaining soil fertility. A majority of these creatures add to the con-tinuity of supply of nitrogenous compounds, phosphorus, and sulfur in the levels reached by roots. As the seventeenth-century English poet George Herbert rightly guessed:

> *More servants wait on man*
> *Than he'll take note of. In every path*
> *He treads down that which doth befriend him.*

When the fertility of a field begins to slip, the temptation is to add inorganic fertilizer, choosing the particular sub-stances the next crop of economic plants will need. Surely the spreading machine can correct in a few hours any de-ficiency in minerals and nitrogen. Theoretically, this is much more efficient than fallowing the field for a year or two, leaving it idle and unproductive. After application of fertilizer, the soil can be put to work again at once.

Unfortunately, the inorganic fertilizers do little to help the microscopic life below the surface. Instead, they may push an invisible succession further in the direction of providing still less nourishment for future crops of harvestable plants. When a period of drought arrives, the land may cease to produce food for man. Or it can become a true desert, perhaps beyond reclamation in one man's lifetime.

In no previous period of history has man known so much about the interwoven forces of nature. Never before has he been able to trace so clearly the successive changes in the life of the land as it adjusts itself to man's activities. Yet we are so close to these new concepts that it is still difficult to view them in perspective.

As we look toward the past, hoping to benefit from human experience, nomadism and subsistence gardens seem the only patterns of land use capable of indefinite futures. Neither of them can support civilization, so that praise for them must be both sincere and faint. Neither may be recommended as a substitute for exploitative agriculture or extractive stock-raising. A new approach is needed. Can the high standard of living on which Thoreau turned his back be maintained if man brings himself into balance with the other animals and plants whose sun and soil he shares?

As Mrs. Peterkin raised the cup of coffee to her lips, she suddenly realized that she had put salt in it, not sugar. She summoned a chemist; he recommended adding a pinch of potassium chlorate. But the coffee tasted no better.

He tried tartaric acid and hypersulfate of lime; these did not help.

In turn he introduced some ammonia, some oxalic, cyanic, acetic, phosphoric, chloric, hyperchloric, sulfuric, boracic, silicic, nitric, formic, nitrous, and carbolic acid. Mrs. Peterkin tested each new mixture and shook her head. It still wasn't coffee.

After another round of unsuccessful experiments with various herbs, Mrs. Peterkin's daughter took the problem to a lady in Philadelphia, who asked: "Why doesn't your mother make a fresh cup of coffee?"

LUCRETIA P. HALE: *The Peterkin Papers* (1880)

CHAPTER · 5

New Patches on Old

ALTHOUGH scientists are busy charting courses to distant planets, the earth has never recovered from the discovery of the New World by explorers from the Old. Plants and animals carried by man from one hemisphere to the other have gone a few at a time, and are still going—keeping nature out of balance almost everywhere.

Many of us tend to forget the early introductions, and to think that today's patterns of living have existed in the same places for millennia. Brazil suggests coffee plantations. Ireland means potato fields. The West Indies imply sugar cane. Yet none of the "coffee republics" had their present source of wealth until coffee was introduced from Africa. The "Irish" potato originated in South America. And sugar cane is an East Indian grass.

Often, as in these instances, the outcome of transplantation has been happy. Other introductions have so changed the course of history that no one can now reconstruct what would have happened had nature been left alone.

Certainly the Jamaica on which Christopher Columbus landed in 1494 was very different from the island of today. Save for the scattered clearings made by Arawak Indians, a great forest of magnificent timber trees extended from the shore to the mountaintops. The Indians depended upon cassava, a starchy food purified with care from the poisonous root of the manioc—a plant they had brought with them in prehistoric times from South and Central America. The tribesmen may have had from the same source small pineapples, ground nuts, and arrowroot. But none of these had much appeal to the hungry Spaniards. Nor were there foresters among the newcomers to delight in the profusion of mahogany, bulletwood, fiddlewood, cedar, and hibiscus-flowered mahoe.

The Spanish colonists brought sugar cane to the West Indies in 1520, a few years after the Portuguese had introduced the plant from the Old World into Brazil. The Spanish made little attempt to cultivate cane, and the first real plantation was laid out in 1660 on Jamaica by Sir Thomas Modyford. It prospered so well that sugar estates spread widely through the islands. For tractable labor, as the Arawaks proved unwilling, Negro slaves were brought from Africa. And the fine forests were cut to make room for cane fields, or to yield fuel with which to boil down the juice of crushed cane and make molasses. These changes in the name of civilization catered to the sweet tooth of the world, and to a growing market for Jamaican rum.

Sugar cane appealed to rats as well. Probably the European brown rat and the black rat had reached the islands aboard the earliest ships. Spiny rats, whose fur is mixed with flattened spines, may have been there when Columbus landed, as animals introduced inadvertently by the canoe-borne Indians. Certainly the cane rat of West Africa arrived aboard the slavers' vessels. Soon all four were thriving, competing for sugar cane.

Rat-catching was a familiar task to European settlers, and soon the planters put a bounty of a penny a head on rats to encourage trapping. The records show that as many as twenty thousand rats a year were destroyed on one Jamaican sugar estate to reap this reward. Rat bounties became a big expense to plantation owners, without making much dent in the number of rodents or in the damage they caused.

Under circumstances of this kind, man casts about for alternative solutions. If his financial garment begins to tear, he looks for a patch to apply. The need for one in the West Indies was far greater than on any mainland plantation because the islands have so few large birds of prey and no predatory mammals. Without wildcats or foxes, weasels or wolves, eagles or big hawks, man seemed to have as allies in his war on rats only a few large snakes, particularly the yellow snake of Jamaica.

The yellow snake is non-poisonous, and hunts at night. It is a constrictor, known to reach a length of nine feet. Reputedly an occasional individual achieves as much as twenty feet. The slaves feared every snake, and killed any they could find by chopping it to pieces with "cutlasses"—those hooked machetes used in harvesting cane. Snakes were effective in ridding Jamaica of rats that ventured beyond the cane plantations, but rarely in killing those that stayed in the cane.

By 1762, barely over a century after cane culture became big business on the islands, the rat problem was so acute that planters were looking elsewhere for new allies in rat control. Sir Thomas Raffles, the English colonial administrator, observed in Cuba that a voracious ant, *Formica omnivora*, attacked young rats while they were helpless in the nest. Swarms of the insects co-operated so well that even the parent rats were seldom able to save their offspring. Sir Thomas had a colony of the Cuban ants dug up and brought to Jamaica, where they settled successfully. Their descendants are known as "Tom Raffle" ants. For a time they seemed effective. But soon they spread beyond the cane fields and became pests in their own right.

In spite of ants, the rat population continued to grow. The English planters thought back to boyhood days in Britain and wondered whether any animal from the Old Country might possibly be helpful in Jamaica. They recalled the intentness with which the weasel-like ferrets had explored the rabbit warrens for victims. How quick they had been! To be sure, the West Indies could offer a ferret no rabbits and few warm-blooded mammals except rats. Surely the introduction of ferrets could do no harm. Perhaps it would solve the rat problem. The planters brought in a number and set them free. But Jamaican chiggers overcame the ferrets before any measurable decrease could be detected among the cane rats, black rats, brown rats, and spiny rats in the cane fields. Ferrets simply died out.

Planters who visited the cane estates on the north coast of South America were decidedly envious when they learned that rats constituted no real problem there. Yet the only conspicuous, meat-eating animals in the fields were huge toads, *Bufo marinus*, with warty bodies as much as eight inches long. A Mr. Anthony Davis introduced these toads on Jamaica in 1844. They throve. They may have eaten some young rats. Certainly they dined on Jamaican insects.

More than a century later we found this spectacular toad still doing well on the islands. It was introduced to us as the "bull frog," and we were told that local dogs either leave the toads alone or die from biting their highly poisonous skin. Today the dogs of southern Florida are becoming acquainted with the same alternatives, for the toad was introduced there during the early 1950's.

As the toads did not work out as hoped, a better patch was needed for the economy. By 1870 each estate was losing about a fifth of its crop to rats, in spite of expenditures running into hundreds of pounds sterling for the services of rat-catchers. In the Swift River Valley some Jamaican estates were abandoned, apparently because losses to rats had become unbearable. The total yearly loss to rats on the island was estimated as not less than £100,000. No wonder the planters were willing to try, as the director of the Institute of Jamaica said, "one darn thing after another!"

The fourth remedy was urged upon W. Bancroft Espeut, Esq., by his wife. She had had a pet mongoose in India as a girl, and recalled that it had killed rats around the house. She showed her husband a statement in the 1816 book *Hortus Jamaicensis* that the mongoose had a natural antipathy to rats. Why not import some? Her husband listened to her, then petitioned the government for permission to import mongooses. After a lengthy correspondence and frequent personal conferences, the license was issued. Mr. Espeut bought four males and five females in Calcutta and released them February 13, 1872, on his Spring Garden Estate in Jamaica. Seemingly, these nine animals were the ancestors of many millions of mongooses in the West Indies today.

Within ten years the planters were complimenting Mrs. Espeut on her recommendation. Mongooses had increased in numbers and attacked the rats so ardently that African cane rats had disappeared, the spiny rats had become extinct

on many islands, the brown Norway rats had ceased to be a menace, and the black rats had saved themselves only by taking to the trees—nesting well above the ground where mongooses did not bother them. In 1882 a census of rat damage was taken in Jamaica, and mongooses were credited with saving the planters nearly £45,000 annually. Abandoned sugar estates were reopened. Mongooses seemed to be the answer. Surely damage to cane would soon shrink into negligibility.

Such a wonderful animal was worth introducing on all other West Indian islands where cane fields were infested with rats. A decade later the deed was done, either with mongooses from Jamaica or more of the same kind from India. This was the same animal Rudyard Kipling made famous as Rikki Tikki Tavi, the cobra-killer, the friend of man. Seemingly, mongooses could do no wrong. And for nearly ninety years now they have been slithering like brown shadows across roads in the cane country, hiding under brush heaps, depending upon speed to escape if man approaches.

On the British island of Dominica the mongoose achieved no foothold. No one is certain today whether the dense forests, or the moisture in this wettest spot of the West Indies, or some other feature prevented colonization. In India itself, mongooses are regarded highly for their ability to kill snakes. Perhaps island snakes were different. Dominica has a boa similar to the constrictor snakes on nearby St. Lucia and Grenada and far-south Trinidad. These snakes climbed through the bushes and low trees and often dropped without warning on mongooses, smothering them in muscular coils. In some of the islands these boas served as a firm control on mongoose numbers, disappointing the cane-planters because of the comparative failure of mongooses to fit quickly into the local scene and thrive.

On the French island of Martinique, as in most of the Lesser Antilles from there to Trinidad, the poisonous fer-de-lance is at home. Residents of these islands assured us that slave-owners had imported the snakes and freed them to discourage desertion by members of the captive labor force. Scientists doubt this story, believing instead that the fer-de-lance was well established in the Lesser Antilles before any kind of man appeared on the scene.

Other residents informed us that the real reason for importing the mongoose onto St. Lucia and Grenada had been to eliminate the fer-de-lance. But each has held a place. Mongooses appear afraid of the quick-striking fer-de-lance, and the great coconut plantations that otherwise offer such inviting territory for mongooses are almost free of the exotic predator. Fer-de-lances hide during the day under piles of fallen leaves, and hunt at night, using their heat-sensitive pit organs between eye and nostril as a means for finding warm-blooded prey in darkness. Mongooses need light to battle a fer-de-lance, and stay principally in the cane fields, where these snakes seldom go. A relative of our whip-poorwill seems to profit from this arrangement. The rufous nightjar rests by day in areas populated by fer-de-lances and thereby is free of attack by mongooses. The bird flies in search of insects when the snakes are active on the ground.

On other islands around the fringe of the Caribbean the mongooses found no obvious competition. They made such inroads on the rat population that rats became harder to find. Mongooses then began ravaging the native mammals, the birds that fed or nested on the ground, the snakes, lizards, amphibians, and land crabs. They even took to eating sugar cane.

Within five years of the introduction of mongooses, the little rice rat of Jamaica—known in no other place on earth —became extinct. Wrens vanished from all islands where they formerly nested in numbers, except on Dominica (where no mongooses got started) and on Grenada (where the local boas kept the newcomers in check). Thrashers and tremblers and ground-nesting doves dwindled to extinction, or their numbers fell to a low ebb from which they have only recently begun to recover because the survivors took to altered habits in nest-building.

Largely because of mongooses, more kinds of animals have been exterminated in the West Indies since 1492 than on the entire continent of North America. Others are so rare that we spent weeks trying to locate survivors. Mongooses seemed to be everywhere, darting across the roads ahead of the car both day and night, or appearing and vanishing when we sat beside a forest trail on the dry side of a mountain to watch for animal life.

The famous yellow snake of Jamaica, most effective of the island's native rat-catchers, disappeared from each region as mongooses moved in. Now it lives only in remote localities. The rare Jamaican coney, a strange little rodent, found safety solely in the moist woodlands of the John Crow Mountains toward the eastern end of the island. The formerly ubiquitous ground lizards, *Ameiva*, dwindled to those on lawns and flower beds in towns and cities. Of ground-feeding creatures, the freest from attack were kinds able to tolerate high humidity or deep forest or the feet of man. Even the giant toads showed some decrease in numbers, for mongooses ate the tadpoles as they emerged from the ponds, their skin not yet fully studded with poison glands.

On many occasions we kept our selves and equipment dry during brief rainstorms in West Indian Dry Season by accepting the gracious invitation of a peasant family to come under the roof of their little house. Almost invariably we found pups and kittens tied up, chicks and ducklings and even lambs in cages well above the ground. This was necessary, the people told us, because any wanderers from the household were almost sure to meet a mongoose and be killed.

Dogs and cats could take care of themselves, as could fully grown hens and sheep. However, the dog and cat were needed in the house to keep down rats and mice. Without them, human habitations became sanctuaries for rodents, not only as "roof rats" but running everywhere, because no mongoose would follow them indoors. Sometimes our own stopping places, when catless and dogless, were unbearably noisy with rats above the ceiling. Bats in the same quarters were silent by comparison.

In Puerto Rico the destruction of *Ameiva* lizards by mongooses led to a new difficulty. White grubs, which are the young of native May beetles there, began to multiply as never before. They undermined the cane, chewed on its roots, and greatly reduced the sugar yield. Adult beetles escaped from the soil into mating flights faster than mongooses could catch them. The former population of *Ameiva* lizards had burrowed extensively after white grubs and kept them in check. Now cane damage from insects rose rapidly, taking the place of rat-caused destruction.

Not until 1920 did anyone realize why Puerto Rico was

afflicted with white grubs while Jamaica, so near and so similar, remained immune. Then someone noticed that Jamaica had the giant toads from South America, and that these amphibians kept May beetles under control. In that year, and again in 1922–3, large numbers of giant toads were imported into Puerto Rico and freed. For a few years the problem seemed solved. The human population grew used to the toads, and Puerto Rican dogs either shied away or died from the poison in the skin of *Bufo marinus*. But some combination of circumstances in Puerto Rico proved unfavorable to the big amphibians, and the toad population gradually shrank. Up rose the May beetles once more. This time the economic entomologists moved in with insecticides. Now the cane-planters had merely exchanged the price of bounties to rat-catchers for bills to chemical-manufacturers. They had the mongooses too.

Seeing the threat on islands where mongooses had become plentiful, the governments on mongoose-free islands began insisting on legal protection, forbidding the import of the potential pest. As far back as 1890 the Governor of Jamaica set up a commission to report on the growing menace from mongooses because serious injury to small landholders and to native wildlife seemed to be outweighing the benefits received by sugar-planters. In a few places, bounties were offered for mongoose heads, just as formerly they had been paid for rats.

In Trinidad the mongoose was too well established for anyone to dream of exterminating the animal. Instead, a food survey was conducted, based on examination of stomach contents. Evidently the wild birds eaten by mongooses about equaled those which would have been eaten by rats if the mongooses had not killed the rats first. The insect-eating lizards devoured by mongooses might or might not have destroyed as many insects as mongooses themselves consumed. The good and the bad seemed about equal, but the report ended with a strong recommendation that no further introductions of this exotic predator be made. How different was this 1918 conclusion from that expressed in 1882 by W. Bancroft Espeut, Esq., in the *Proceedings of the Zoological Society of London:*

> I question much if such enormous benefit has ever resulted from the introduction and acclimatization of any one animal, as that which has attended the Mungoos in Jamaica and the West Indies; and I marvel that Australia and New Zealand do not obtain this useful animal in order to destroy the plague of Rabbits in those countries.

In 1950 a new study of the mongooses in the West Indies was made, to learn how they were faring now that they had reached an ecological balance with their environment. On the Virgin Islands these animals were found to depend primarily on insects, with about equal parts of harmless amphibians, mice, rats, and crabs, and smaller amounts of fruit and vegetable matter. More mongoose stomachs included sugar cane than held either poultry or lizards. But this was taken to indicate how thoroughly the mongooses had eliminated the available lizards, and how well the islanders are now protecting their chickens.

Simultaneously a census of rats was taken. Arboreal black rats had increased in numbers, in some places to as many as twenty to the acre. Damage to the cane crop from this original pest stood between an eighth and a quarter of the total stand. Even with giant toads to help them, the planters were having to spend more on insecticides than ever before. Rat damage about equaled that in 1872 before mongooses were introduced.

The world price for sugar might never have justified any positive action against mongooses in the West Indies, and the smaller landholders combined could scarcely induce the government to spend large sums in aid. The *status quo* seemed likely to remain unchanged indefinitely. Suddenly, from public-health officers came evidence that mongooses constitute a reservoir of rabies. Occasionally a domestic dog was bitten by a rabid mongoose, contracted the dread disease, and passed it along to man. In Puerto Rico this new discovery led to studied attempts to cut down on the mongoose population, even if rats and insects increased in numbers as a consequence.

Mongooses will eat almost anything, including carrion. Sun-dried fish proved to be an effective bait that could be drugged with thallium sulfate or with the famous "1080" poison so useful in Australia and New Zealand in the recently stepped-up war on introduced European rabbits. At concentrations so low that the mongooses could not detect the poison in the fish, thallium sulfate had a cumulative effect. Three meals of it killed a mongoose. On Pineros Island, just off Puerto Rico, all adult mongooses apparently were killed by this method. On Puerto Rico itself, a ninety-per-cent reduction in mongooses was obtained at a cost of about seventeen cents per acre.

It is still too early to know whether this gain on mongooses can be maintained or improved upon, or to guess how the rats, mice, crabs, and amphibians will fare if mongooses are reduced to the ten-per-cent level. Jamaica's yellow snake might return to the cane fields, along with *Ameiva* lizards from the city dooryards where mongooses have left them alone. The coneys might spread from their strongholds in the John Crow Mountains. Animals exterminated by the mongooses on some islands might be reintroduced from others. Kinds now extinct everywhere are gone, of course, forever.

Any continued success against mongooses in the West Indies would be followed, almost certainly, by similar measures on Hawaiian islands, where the same animal was introduced in 1883 with comparable results. The marvel is that as recently as 1951 a shipment of eighty mongooses went legally from Jamaica to Colombia for release as a measure against snakes there. Previously they had been freed in South America only in British Guiana, where they spread in the coastal cane country, but were barred from the interior and from the rest of South America by the dense rain forest. Until the Colombian introduction, scientists had hoped that mongooses could be exterminated in the Guianas before the jungle barrier was felled. But now the future is not so bright.

Our own United States Department of Agriculture once gave serious consideration to introducing the mongoose into western states as a control for pocket gophers. This was in 1898, when enough people had already had sad experiences with mongooses in Jamaica and in Hawaii to nip the idea in the bud. And today the newspapers include an occasional item describing the plight of the pocket gophers. In Kansas and elsewhere, both the gophers and the prairie dogs seem in danger of becoming extinct wherever farmers and ranchers prevent overgrazing and let the grass grow tall. So simple a change, with only the native predators present, can mean success or failure for these rodents no mongoose has ever chased.

Foresight, based on experience elsewhere, saved America from mongooses. A local situation did not turn into a national blight. The West Indies and Hawaii suffered from lack of precedent, as people did not recognize how little the rats and the cane affected the rest of island life. Each new patch for the sugar economy disturbed the balance of nature far more than the previous one, and brought a need for still more patches. The pattern of the crazy quilt soon passed beyond anyone's understanding. How much better the island life would have fared had the original balance been left alone!

THE crocodile is an animal placed at a happy distance from the inhabitants of Europe, and formidable only in those regions where men are scarce, and arts are but little known. In all the cultivated and populated parts of the world, the great animals are entirely banished, or rarely seen. The appearance of such raises at once a whole country up in arms to oppose their force; and their lives generally pay the forfeit of their temerity. The crocodile, therefore, that was once so terrible along the banks of the river Nile, is now neither so large, nor its numbers so great as formerly.

OLIVER GOLDSMITH: *A History of the Earth and Animated Nature* (1774)

CHAPTER · 6

What Good Is a Crocodile?

E<small>VERY</small> so often someone asks "What good is a mosquito?" Or a wolf? Or a crocodile? The question, deriving from the common antipathy to these creatures, is largely rhetorical. Any reply other than "None whatever!" is likely to fall on closed ears. Usually emotions block any open-mindedness or acceptance of facts regarding predators and parasites.

The "good" is always in terms of usefulness to mankind, of course, as though democratic principles could never be extended to other kinds of life, each with an equal right to thrive on earth. A mosquito is the ideal mate for another mosquito, and crocodiles do everything they can to ensure the biological success of crocodiles.

Most people would be happy to see mosquitoes exterminated. Many feel the same way about crocodiles, and would include for good measure all other crocodilians—the alligators of America and the Orient, the caimans of Middle and South America, and the gavials of India, Sumatra, and Borneo.

In Africa the Nile crocodile—a widespread kind—is credited as the number-two killer of human beings, second only to poisonous snakes. In populous India, cobras and other venomous reptiles cause the death of more than sixteen thousand people annually, yet are less feared than the crocodiles and gavials, which kill only about 250. Perhaps the vegetarian Hindus are particularly horrified to realize that crocodiles usually eat their victims, whereas snakes merely strike and slither off.

The Western Hemisphere has crocodiles as well as alligators and caimans, but few records of human deaths from any of them. Unlike the slow-moving and seemingly clumsy alligator with its broadly rounded muzzle, the narrow-snouted crocodile is alert, agile, and swift. So wary is it that, unless the wheather is chilly, a person has difficulty getting a close view of the animal in the wild. But if the water temperature falls as low as sixty-five degrees Fahrenheit, crocodiles become so torpid that they may drown even in shallows. The American alligator, by contrast, tolerates any weather above the freezing point and, at the approach of winter, habitually gorges itself before hibernating for a month or two.

American crocodiles prowl through the swamps of the Tropics as far south as Colombia and Ecuador, competing with alligators over some of this range. Crocs inhabit the mangrove tangles along the Florida keys, taking advantage of the consistent warmth guaranteed by the adjacent Gulf Stream. Near Cape Canaveral, at the head of the natural intracoastal waterway known as the "Indian River," is the northernmost point any crocodile has been known to reach. There, in 1872, the pioneer C. J. Maynard killed a 10-footer close to Lake Harney, in what is now Volusia County. Probably the animal had followed the waterway, traveling from the keys in a series of warm years. Alligators, by contrast, are known all the way to the coastal Carolinas, as well as around the Gulf of Mexico and for some distance up the Mississippi. Both of these crocodilians have been caught on the major islands from Cuba to Puerto Rico.

The largest American crocodile for which an authentic record is available measured twenty-three feet, having reached this length in South America. But the crocs of Madagascar, like the Indian gavials, may reach a length of thirty feet—only slightly shorter than the longest snakes. Moreover, they run as though on tiptoes, with body well above the ground, then lash out expertly with the tail—often sweeping a victim off its feet, into reach of the tooth-studded jaws. Crocodilians should be treated with respect!

After watching a full-grown croc in action, a person can be far more grateful that the carnivorous dinosaurs are all extinct—particularly the giant forty-seven foot *Tyrannosaurus*. One crocodile fossil from Texas is forty-five feet long, with jaws gaping for a length of three feet. This not too different ancestor was a contemporary of *Tyrannosaurus*.

Fishermen in tropical waters have long been the enemies of crocodiles and alligators, for these animals get caught in the nets and tear them while robbing them of fish. Whether the reptiles act as scavengers or depend significantly on a fish diet was rarely questioned until recently. Most of us seem convinced on the subject because we were brought up with

57

the notion, if only from Lewis Carroll's rhyme in *Alice in Wonderland*:

> *How doth the little crocodile*
> *Improve his shining tail*
> *And pour the waters of the Nile*
> *On every golden scale!*
>
> *How cheerfully he seems to grin,*
> *How neatly spread his claws,*
> *And welcomes little fishes in*
> *With gently smiling jaws!*

In estimating the damage done by crocodiles to their business, fishermen tend to consider the big reptiles as exclusively fish-eaters, and as having demands for flesh equivalent to those of a carnivorous mammal of corresponding weight. If these assumptions were correct, a 750-pound croc probably would require, as J. F. E. Bloss claimed, "as much fish as a hundred men, and surely take more fish from the river than the tribesmen do." This claim in the *Sudan Notes and Records* is not borne out by fact, for crocs select a varied diet and lead such leisurely, cold-blooded lives that their food needs are actually modest.

Crocodilians do eat fish. Yet, when the reptiles are young, the tables are often turned. Many little crocs venture over deeper water, even out to sea, where large predatory fish can reach them. Egrets and herons, which normally spear fish, ease their hunger by swallowing small alligators and probably young crocs as well. It is only after evading such hazards year after year that any crocodilian can reach a length of several feet and some degree of immunity from non-human enemies. Then each becomes a target for the hide-hunter, greatest danger of all.

Both fish and crocodile, as stuffed skins, were favorite ornaments in scientific laboratories of the seventeenth century. Old paintings, etchings, and engravings almost invariably show one or more of each, hung from the ceiling or attached to the walls where alchemists experimented. The fish was an early symbol of Christ, and was believed to help create a favorable environment in which to transmute base metals into gold. The crocodile represented health, the realm of the physician, because of its supposed virtues in the treatment of human ills.

Between 1600 and 1800 medical practice was so little beyond the level of the witch doctor that crocodile parts were in great demand. The blood, usually dried, was a sovereign remedy for snake bite, as well as for treatment of eye disorders. The bile and gall bladder, similarly desiccated and pulverized, were valued ingredients of salves and ointments intended to improve vision. Crocodile fat became a special lubricant for massages intended to drive out fevers. The animal's hide, when dry and powdered, was mixed with oil or vinegar as an anesthetic on parts of the body about to be lanced or cauterized or amputated. Crocodile flesh, commonly fried, was recommended for removing the pain from a wound. Even the reptile's dung supposedly had wondrous powers to induce renewed hair growth on bald men—a fertilizer with magic in it!

No one knows the routes taken by these beliefs from lands inhabited by crocodilians into Europe during the Renaissance. It is much easier to see how the widespread use of crocodile teeth in Africa as charms to ward off bad luck, counteract poison, and prevent disease is connected with the continued inclusion of very similar alligator teeth in the "conjure bag" used by voodoo adherents among southern Negroes today. Seminole Indians, and others of their race in regions where gators live, may have adopted from Negro slaves the idea of wearing a crocodilian tooth on a cord around the neck.

All of these values are in dead, not live, crocodilians.

Only a tourist seems able to find a thrill in seeing a big gator basking beside a walkway in the Everglades or in the great Okefinokee Swamp between Florida and Georgia. Or a group of crocs facing the river from a bare mud area along an African waterway. Almost no one else has a kind word for monster reptiles.

Steadily the habitat of crocodilians the world over is being reduced through drainage of swamps, hydroelectric schemes, and development of land for human communities. Native guides and white hunters still lead the quivering sportsmen from Europe and America on safaris to vantage points from which their high-powered rifles can reach motionless crocodiles whose wet armor plates gleam in the sunlight.

During the past twenty years the market for crocodile and alligator hides grew enormously. Thousands of skins monthly were needed to supply the world demand for shoes and ladies' purses. In our southeastern states, men poled boats quietly at night along waterways in the great swamps, shining lights that might be reflected by "Ol' Fire Eyes"—a big gator with a valuable skin. In Africa, natives to whom money had come to have a meaning grew bold in broad daylight. With spears poised, they drifted in dugouts to where a croc floated like a log, only its nostrils and eyes above the surface.

So many crocodile carcasses were wasted—left for the vultures to clean—that a few people wondered whether some better use could not be found for them. Volume XV of the *East African Agricultural Journal*, published in 1950, contains an account of "A Preliminary Test of Dried Crocodile Meat in the Feeding of Pigs," with limited data showing that the waste meat was quite usable. Pigs fed on it grew as rapidly and to as high quality as those fed on commercial meat meal. Seemingly, no one thought of canning crocodile meat, as is done with rattlesnake filets, for sale in luxury delicatessens.

Suddenly in the 1940's the fortunes of crocodilians all over the world took a decided turn for the worse. In Florida the growing scarcity of gators alarmed both the hide-hunters and the state Game and Fresh Water Fish Commission. Laws were introduced and passed in 1944 to protect all nests, young, and adults of both the alligator and the crocodile in Florida. In 1947, Royal Palm State Park was expanded into the Everglades National Park to protect a large area of habitat as well.

Florida law forbids the shipment of hides and crocodilians alike. No longer can foot-long gators be sent alive to the "folks back home" in New England as potential pets. Nor can small stuffed bodies be offered for sale as curios to gather dust on mantelpieces. Still the demand for skins continues, and tourists seek the souvenirs to which they have grown accustomed.

Displaced Yankee ingenuity found a way to supply tourist markets without running afoul of the law. Skillful collectors fine-combed Central and South American swamps to catch thousands of young spectacled caimans for shipment to Florida and sale as "baby alligators." Caimans are gatorlike in the roundness of their snouts, and only an expert is likely to notice the bony brow ridge extending across the head from eye to eye, or the hard feel of the skin due to underlying thin, bone-like plates. Game and Fish Commission inspectors know these differences, and with one grasp around the body of a "baby alligator" can tell whether it is soft and real or firm and imported.

These distinctions become academic north of Mason and Dixon's line when the express agent brings a live crocodilian as a gift to some unsuspecting family. Then our telephone is likely to ring. Excitedly the party at the other end of the line fires questions: "What do I feed it? How long will it live? How big will it grow?"

Fortunately, caimans usually adapt themselves quickly to captivity, just as do most gators. Soon they may accept earthworms, insects, even proffered bits of hamburger or horsemeat. But rarely does one reach a foster home so sympathetic, tolerant, and hospitable that the crocodilian will be tended year after year and grow to full size—twenty feet for a caiman. Instead, after a few months the barely larger "baby" dies if it is not given to a zoo or school for public display.

Slowly the American gator has recovered in numbers. Since 1950, Florida has tried an open season on gators eight or more feet long from October 1 through January 31 in northern counties of the state. Special permits are sold, and just ten days are allowed following the close of the season for legal disposal of hides and for their shipment out of the state. Whether gators become sexually mature before reaching a length of nine feet has never been settled, but the tight control seems to be working out.

In both Florida and California, farms operated like fish hatcheries are now raising crocodilians. In large pens with pools, the reptiles are fed diets calculated to induce rapid growth to a legal size. Then each animal is sacrificed to become a hide for the market. Crocs do not do as well as gators, yet millions of tourists pay to see the reptiles. Often their admissions fully cover the cost of the undertaking. To answer the endless questions, the custodians are accumulating a considerable knowledge on the natural history of the animals.

What a long way human thinking has come since the days of the Pharaohs! Then the Nile crocodile was a sacred beast, dedicated to Sebek, the deity of evil, representing the destroying power of the sun. Giant crocs served the Egyptians both in theology and refuse disposal. Even unwanted children were thrown to the crocodiles, simultaneously propitiating Sebek and keeping the human population nearer the limits imposed by its food supply.

With the remaining stock of crocodilians shrinking over much of the tropical world, Madagascar and Guatemala have already introduced legislation to protect the survivors. The degree of enforcement may still be inadequate to save them. Elsewhere the situation is changing so rapidly that naturalists are being invited to study crocodiles before it is too late.

The British government in 1952 helped the eminent English scientist Dr. Hugh B. Cott to look into the decline of crocodiles from Rhodesia to Uganda. In a short time he got a far more adequate answer than did Kipling's delightful Elephant's Child to the question: "What does the Crocodile have for dinner?" Dr. Cott learned how complex are the ways the big reptile fits its environment, and how a reduction in crocodile numbers is altering that part of Africa.

Crocodiles are unlike alligators in that they build no nest of mud and plant material. Instead, the female croc scoops out a hole in sand or in the soil just back of a beach. She makes a cavity about a foot deep and eighteen inches in diameter as a repository for some two dozen hard-shelled eggs, each about the size of a goose egg. She covers them with sand and tamps the surface until it is smooth. Then, except for occasional visits as though for inspection, she leaves the eggs to hatch on their own. Only if she hears the raucous, high-pitched "distress call" of hatchling crocs will she come charging to defend them.

Young African crocodiles shun the open waters, apparently because the adults there are ready for cannibalism at any time. Little crocs stay in weedy areas, often scrambling like lizards among the bushes in search of insects and spiders, their chief food at this stage. Dr. Cott hid in the equivalent of a bird blind from dawn until dusk and saw that they take also carnivorous water beetles, the aquatic young of dragonflies, vast numbers of small crabs, and many other pond or stream denizens that feed extensively on young fish. In this way the presence of crocodiles actually supports the fishing industry.

As the croc grows, it seeks larger prey among the frogs, toads, crabs, and snails available in the tangle of papyrus stems and shore vegetation. Occasionally it may catch a small bird or a fish, even an unwary rodent of some kind. Gradually its diet changes. The croc becomes more aquatic and nocturnal. The pupils of its eyes open wide while the animal prowls at night, but close to mere vertical slits like those of a cat while it basks by day.

Through adolescence, to an age of eight or ten years, crocodiles subsist chiefly on fish. They specialize, however, on slow-moving scavengers such as catfish and carnivores that work the bottom near shore in search of fish eggs and fry. In this way the crocs protect fish the fishermen want. In fact, commercially valuable fish proved rare in crocodile stomachs, regardless of the time of day or year when the reptile was examined.

With continued growth the crocodile's voice changes. While its body is between four and five feet long, the baby "distress call" disappears and is replaced by a rarely used roar, a bellow like a deep-throated hiss amplified a thousand times. Simultaneously the diet alters again to include less fish and more fish-eating birds, more mongooses, more otters, and a variety of shore visitors that venture too close while coming for a drink in the dark.

Really big crocodiles seem to eat less frequently, and often go for days with an empty stomach. Yet they also succeed in dining on a wide range of food: monitor lizards, young of their own kind, snakes (including cobras and giant pythons), soft-shell turtles, ducks and other waterfowl, careless antelopes, buffalo, hippopotamus calves, and the

occasional man who falls from a canoe or child who wanders within reach. A duck may be swallowed whole, but larger prey is snapped into pieces first. Often the crocodile gasps for air as a big morsel goes down its throat, and at these times its tear glands discharge copiously, furnishing the basis for the stories about "crocodile tears" and the anthropomorphism that these reptiles weep in remorse while engulfing their victims.

After a meal, crocodiles commonly rest on the riverbank with mouths agape—a position rarely seen in alligators or caimans. From this habit came the incorrect idea repeated by Herodotus, the Father of History, in the fifth century B.C., that the upper jaw of a crocodile is the hinged one, not the lower as in all other animals. While its mouth is open, the croc permits spur-winged plovers to run freely over its tongue, like animated toothbrushes picking out shreds of meat. Perhaps the croc can take an after-dinner snooze as long as these alert, noisy companions are near. Certainly the big reptiles seem apprehensive when the birds cry out and fly away.

One reason for the suddenness of the change in crocodile numbers is the relationship long maintained with the semi-aquatic monitor lizards. The monitor *Varanus niloticus* is a smaller relative of the ten-foot "dragons" of the Lesser Sunda Islands in the East Indies. For millennia these monitors have been respected and represented in picture-writings of ancient Egypt because of their habit of eating the eggs of crocodiles.

So long as adult crocs were abundant, they kept monitors under control. Crocodile nests were vigorously defended by the returning parent on her "inspection trips," and the proportion of young destroyed by monitors seemed negligible. But as crocodiles declined, monitor lizards throve. Now crocodile nests are too few to withstand such persistent raiding, and monitors freely patrol the watercourses, hunting young crocs as well. By killing off the mature crocodiles, man has doubly interfered with the animals' reproduction—directly, and indirectly by aiding their principal enemy, the monitor.

Abruptly, Africa has found a new devil to pay! With fewer young crocs policing the shallows, a far larger number of dragonfly young, carnivorous water beetles, and crabs survive. They prey on fish fry, bringing about a serious shortage in the supply of *Tilapia*, an important fish depended upon for proteins in native diets. In some rivers, now that adolescent crocs give little competition, the widespread Mozambique catfish has multiplied beyond anything on record and is devouring even more *Tilapia*—at larger sizes. Both fishermen and villagers are distressed.

As recently as 1945 the fishermen around the southern end of Lake Victoria were complaining bitterly that crocodiles, then abundant, persistently hunted out the African lungfish *Protopterus*. Many natives relished *Protopterus* as a delicacy, partly because it was scarce. Now lungfish are abundant, and no longer a delicacy. They are attacking fishes of more valuable kinds while these are held in nets set by the fishermen. Soon the villagers may have to eat lungish if they want fish at all.

In British East Africa other fishermen are equally distraught because fish in their nets are being destroyed by otters, which have become numerous and intrepid since their chief enemy, the crocodile, became rare. How often do African fishermen think back to the days when they protested over the fish supposedly eaten by crocodiles, and urged elimination of the armored reptiles?

Madagascar has different troubles. There crocs were virtually eliminated by the hide-hunters, and soon the number of stray dogs and wild pigs rose spectacularly. Farmers now demand governmental protection for their crops. Otherwise, they claim, they face financial ruin as wild pigs repeatedly raid cultivated land. And with the increase in stray dogs has come one outbreak of rabies after another. No longer are these animals kept in check by crocs lurking in the shadows by the streams.

Over much of Africa the crocodile is still classed as vermin, to be eliminated in every way possible. At the same time, modern medicine is being called upon to save lives that otherwise would be lost to an increasing number of bites from cobras and rabid dogs. Costly drives against catfish and otters and lungfish are being considered to protect the dwindling supply of food fish—an expensive service the crocodiles in former years performed for free. Crabs are still too elusive, and so are the less-known fish that feed on eggs and fry of *Tilapia*.

Through all this confusion the hide-hunter watches for opportunities to continue and extend his exploitation. He encourages the newspapers, like native gossips, to echo all over Africa every time a child or man or ox is killed by a crocodile. He points with pride to the decrease in human deaths from crocs along the rivers and coasts, and claims honor as a savior—without mentioning the wealth he has gained from the sale of crocodile skins. By day he may worry over the decline in their numbers. But by night he goes jacklighting for immature crocs, killing them long before they can reproduce. Are these crocodile tears he sheds?

It is no news when a croc eats an otter, a lungfish, a cobra, or a rabid dog. But everyone hears about a village that has been flooded because vegetation has clogged a river. This, of course, is merely plant growth. Perhaps its increase is related to a change in the weather, warmer or colder or wetter. The press and the village elders urge the government to have a path cleared for drainage. No one seems to remember that formerly the crocodiles kept the watercourses open and let the river through.

In former years it was no news when a crocodile killed a hippopotamus calf for dinner. Or when the mother hippo rushed to the scene and crushed the armored reptile in her cavernous mouth. But without crocodiles preying on their calves, the vegetarian hippos increased in numbers and grew progressively more destructive. They raided farms near the rivers. Villagers complained with just cause, and hunters were sent in. A modern rifle can place a bullet in a vulnerable spot, providing the village with a brief abundance of hippo meat. Soon the original reason for shooting hippos was forgotten. The hunter's job changed into one of providing the villagers with meat. Gradually the number of hippos shrank far below that at which they had been in equilibrium with abundant crocodiles and the puny weapons of primitive man.

No one seemed to care much about hippos, even if they disappeared entirely. Why save any for the future as a tourist attraction when African natives could benefit today by eating the fresh meat? But the destruction of hippos brought unexpected, unwanted changes. The scattered remainder could no longer maintain a network of trails through the papyrus swamps. The coarse grass grew densely until it served as a broad dam holding back rainwater whenever it fell in quantity. Now farms and communities that suffered from too many hippos are being flooded repeatedly because there are too few. Engineering aid is demanded to correct a situation the hippopotamus population could maintain if numbers were regulated as they used to be by crocodiles.

It may be time to wonder whether leather articles made of crocodile and alligator hide should follow egret feathers on ladies' hats as forbidden possessions. So drastic a step seems necessary to perpetuate a kind of reptile life which has survived virtually unchanged for a hundred million years. Under present rules, far too many covetous eyes are turned toward the crocodiles in Uganda's Murchison Falls National Park and similar sanctuaries set aside for the protection of all kinds of wildlife. Unless the present lucrative market for hides collapses, crocodilians the world over may not survive until nationalism in independent tropical countries comes to include prideful protection of native animals as well.

The earth has many kinds of animals that are appreciated alive no more than the crocodiles. Yet their disappearance would rend the food webs into which they fit, and introduce complications of which man is now blissfully unaware.

I now *suspect that just as a deer herd lives in mortal fear
of its wolves, so does a mountain live in mortal fear of its
deer. And perhaps with better cause, for while a buck pulled
down by wolves can be replaced in two or three years, a
range pulled down by too many deer may fail of replacement in as many decades.*

*The cowman who cleans his range of wolves does not
realize that he is taking over the wolf's job of trimming the
herd to fit the range. He has not learned to think like a
mountain.*

ALDO LEOPOLD: *A Sand County Almanac* (1949)

CHAPTER · 7

A Yardstick for Deer

❁

Deer, unlike the many wild things that come to our
feeding trays when they are hungry, tend to remain invisible. Except for an occasional glimpse or a footprint or a
"sign," we could forget that deer share the fields and woodlands with us—until the hunting season. Then, as the broad-
leaved trees blaze with colors more vivid than a hunter's
cap, we meet deer on every highway, lashed conspicuously
to automobiles. And we know that others have killed them,
harvesting animals that have become weeds.

Deer are so retiring that science tells little about how
much countryside each one needs: what food and cover this
must include, how the seasons modify these requisites, or

even whether the territory used by a deer is determined individually rather than on a family basis. How big is home
for a deer? As with the three princes of Serendip, the answer
turned up where it was not sought.

In the prosperous 1920's Colonel Edwin S. George
bought a country estate about twenty-five miles northwest
of Ann Arbor, Michigan. After his new house was in order,
he decided that his pleasure would be increased if occasionally a deer walked by—wild, and relatively free. The
Colonel's interest was neither in making pets of game animals nor in hunting.

To gratify this urge, a seven-foot fence was ordered and
erected around the twelve-hundred-acre estate. The lower
six feet were woven mesh, and the top was a five-strand
overhang of barbed wire. The fence might not be completely deer-proof, but it would do. Inside the enclosure
Colonel George released four full-grown does and two
bucks from Grand Island, Lake Superior. The white-tails
had no enemies worth mentioning, and from the time of
their arrival in March 1928 they proceeded to enliven the
property just as the owner had planned.

After the stock market crashed, Colonel George donated
the whole estate and its deer to the University of Michigan,
to be administered by the Museum of Zoology as a natural
area. Almost half of the land was grassy, a quarter in mixed
hardwoods (mostly oak and hickory), better than 10 per
cent a tamarack swamp, and the rest smaller proportions of
marsh, brush, and bog. Although the gift was made in 1930,
the university scientists did not realize until the fall of 1933
how phenomenally the deer herd had increased, or the degree to which it was overusing its food supply.

A rough count of deer on December 3, 1933, revealed
160 animals. The professors decided that this was too many,
but could not guess by how much. For a trial, six bucks and
three does were shot in the following month to bring the

population toward 150. An additional animal found dead during the summer of 1934 completed the planned "removal." But that fall a second census revealed 210 deer and still more extensive damage to trees and shrubs. The deer were reproducing like the guinea pigs in Ellis Parker Butler's classic. At least sixty fawns must have been dropped and reared during 1934—an increase of forty per cent in a single season!

Experienced deer-hunters shook their heads at these numbers. No herd could grow so fast. According to them, a doe had to be two years old to bear her first young. Each June thereafter she might drop one, or sometimes two fawns. In any given year, probably ten per cent of the does would bear none. Forty-per-cent increases just couldn't be real!

The professors checked their figures. True, the seven-foot fence had not proved so perfect a barrier as Colonel George had hoped. During the year, two deer had leaped it into the overpopulated reserve, as though they wanted company. But two more had been seen escaping by the same route, heading for the comparative solitude and more abundant food beyond the fence. If deer could escape, the forty-per-cent increase among those remaining captive might represent only part of the actual growth of the herd.

Whether a deer herd could or could not add forty per cent to its numbers in a single year was far less certain than that the George Reserve had too many deer. They destroyed hickory and maple seedlings almost as soon as each germinated. Oaks made slightly better progress, but red osier dogwood, a favorite deer food, seldom got started inside the fence from the thousands of seeds dropped by birds. Beyond the wire, the dogwood formed a conspicuous part of the undergrowth.

The characteristic "browse line" of overpopulated game ranges was already visible on the university's trees, like a soup-bowl haircut. Low branches had all been trimmed by the hungry deer, and shrubs showed extensive damage. The custodians decided on a drastic cut in the herd—ninety-five instead of the mere nine in the preceding year.

Another decision seemed called for on rational grounds. Presumably both bucks and does ate about the same amount of forage, yet does alone produced fawns. Why favor the does by shooting twice as many bucks? A proportion of three to two was agreed upon, and a paid hunter went to work. He took out fifty-eight males and thirty-seven females, a number of each being fawns in their first winter. Each doe, whether mature or not, received careful study.

One surprise came promptly. Some doe fawns less than six months old were pregnant! The lore of the deer-hunters had been incomplete. Years later Drs. Glenn H. Morton and E. L. Cheatum confirmed the discovery in southern New York state. Probably more than a third of fawn-age does give birth to single offspring the following June. Moreover, well above half of the older does may produce twins, rather than a single young.

The George Reserve of the University of Michigan has become a yardstick to measure the deer population an area can support without losing its plant cover. Always the amount of winter food determines the actual number, and this varies somewhat from year to year according to weather conditions in the preceding summer. At first the custodians aimed at a midwinter herd of just above one hundred animals. Then they tried seventy-five, and still more recently fifty-five. Each time they saw progress in protecting the future of the reserve.

Now that a hungry deer is no longer waiting to eat each new leaf as it unfolds, the understory of the forest area has recovered considerably. Paralleling this improvement is an increase in gray squirrels. Shrubs and herbs are spreading. With places to hide, chipmunks and cottontails are moving in. Apparently all the orchids and trilliums and lilies on the reserve had not been destroyed by the deer during the years of overpopulation, for these plants too are beginning to show again.

Toward Christmas each year a new census of the herd is made to calculate the number of fawns dropped the preceding spring, and to decide how many deer should be removed to keep the population at the desired level. Seldom can the custodians predict from the summer just past or the winter before it whether the deer have reproduced well or poorly. Far wider fluctuations are found than anyone had anticipated, but the old figure of a forty-per-cent annual increase is now regarded in another light. With the herd more nearly in balance with its food supply, the reproduction rate has risen; the average stands just under fifty-four per cent. Nor is this a measure of what the deer can actually achieve in a truly good year, for after the herd had been cut back to forty-nine animals one winter, fifty-one fawns appeared the following spring—an increase of 104.1 per cent!

The scientists in charge of the George Reserve have learned how many deer the tract can support if the land is expected to yield nothing else. But the animals still need so much food that pine plantations or farm crops would stand no chance inside the fence. If deer are to live at peace with man in the forests and on the farms he uses for raising trees and vegetables, far fewer can be tolerated. Less than half the number maintained on the reserve may be the limit, with each deer allotted forty to fifty acres. If this ideal is to be realized, one deer for every three in an area must be removed annually without fail, before the severest days of winter come. This is the herculean task today's hunters face.

Whether we like it or not, deer are part of our heritage. Today America has far more of them than when the early colonists came. Deer are creatures of the forest edge, and the country's timberlands have been so dissected by roads and farms and communities that forest edge is abundant. Yet the deer that invades a field affects a farmer or a herdsman. The deer that stays among the trees interferes with their reproduction. Both timber and farm products are too precious to be endangered by surplus deer. How can the deer be used, and at the same time be maintained in balance with man's land?

No hunter can be expected to make an annual census of the deer where he wants to hunt, and democratically to agree with his companions on how many animals should be taken from the nation's living resource. We have entrusted these duties and judgments to state game commissions and the federal Fish and Wildlife Service.

Making a census of game animals is a tremendous undertaking. To place a summary before the public takes time: the tally of big game for 1956 appeared in March 1958. It included nearly eight and a half million deer. If these animals were evenly distributed through the country's forests and woodlands, each deer would have about sixty-one acres to itself. But the herds are in clumps, most of them outstripping the food available.

Before the white man came to America, deer were kept in check by cougars, wolves, and Indians. Perhaps smaller carnivores aided in maintaining a balance between deer and vegetation, but the lynxes, bobcats, foxes, and coyotes were really only nature's second line of defense. Now that man has eliminated virtually all of these restraints on deer reproduction over most of the country, he is obliged to take their place. Yet so much effort seems required to accomplish the task that he cannot refrain from wondering occasionally how the predaceous mammals managed before.

The principal governor on deer was the cougar, known variously as a catamount, an American lion, a mountain lion, a silver lion, a painter, a purple panther, a puma, and a brown tiger. All of these apply to *Felis concolor*, a small-headed, long-tailed cat reaching a body length of nine feet and a

75

weight of two hundred pounds. The cougar specializes in leaping from low limbs in the forest upon a deer passing below, and ordinarily kills a victim on alternate nights to feast on venison. Seven or eight pounds constitutes a full meal, and the rest is covered with leaves—actually abandoned to the many carrion-eaters that gather around.

The pioneers came to expect one of these big cats on every major hill. Even then the number of cougars was not large. Each defended a territory about nine miles across. Within this hunting preserve it stalked or crouched in wait. But when white man introduced horses, trouble began. Cougars preferred horse meat to venison, or perhaps found the domesticated animal easier prey.

Occasionally a cougar develops a liking for domestic cattle, just as a few Indian tigers acquire a taste for human flesh. Usually the offender in both cases is found to be an old cat with waning vigor or a cripple unable to subdue more usual prey. But cattlemen generalize in condemning all cougars and throw their financial and moral weight against any move to discontinue paying bounties for dead cougars. They remain unconvinced that today, since the replacement of the horse by the automobile and tractor, the big cats have gone back almost exclusively to deer. In consequence, thousands of dollars go into the pockets of professional cougar-hunters each year. And crop insurance pays out thousands more for damage done by the deer the bountied cougars didn't kill.

A century of persecution has not displaced cougars from a large proportion of western hills. As long as California's mountain forests remain, clothing some of America's most rugged terrain, the big cats there seem safe. About 150 are shot each year for the reward of fifty dollars each, without protest from the lumbermen and ranchers who complain about the damage done by deer. The same 150 cougars would have killed close to twenty-seven thousand of California's excess deer, and the carrion from discarded carcasses would have supported large numbers of bobcats, coyotes, and foxes. Each of these smaller carnivores could have helped still more in reducing the number of deer fawns reaching reproductive maturity.

In the Old Days a deer the cougar missed was often run down by a pack of wolves. Deer and elk were the favorite foods of the gray wolf until man brought in cattle and sheep. Then these wild dogs transferred their attention and had to be exterminated. Today only a few remain in the United States, either in national parks or hunting through the great forest tracts of north central states such as Minnesota and Wisconsin. In Canada, where the same animal is known more commonly as the timber wolf, it has had less persecution and may be scarcely reduced from pioneer times.

It is possible, of course, that Canadian wolves regularly spread southward into our northern forests. This may be why hunters have been able to kill an average of one wolf per forty-one square miles in the Superior National Forest of Minnesota. The population remains steady at about three or four wolves to the area a cougar would defend. In earlier centuries comparable numbers may have been present over much of the continent, chasing deer by day in the same regions where cougars prowled at night.

Ordinarily, gray wolves work in groups. In open country they may trot tirelessy all day, keeping a deer running until it is exhausted. Where the terrain is rougher and dotted with lakes or swamps, the carnivores usually act in relays, driving a deer from one wolf's station to the next until the victim becomes so confused that a member of the team can bring it down.

Wolves tend to gorge themselves on the site of the kill, eating as much as a fifth of their own weight in one sitting. If the prey is large, each wolf may dig a hole, disgorge into it a first meal as a cache to be consumed later, then return to the prey for an equally generous refill. In this way a family

77

group, consisting of a 150-pound adult male, his 80-pound mate, and two 75-pound youngsters, can easily dispose of 150 pounds of meat. Ordinarily one deer sustains the four wolves for a week, keeping them so satisfied that they will not kill another. They leave little to attract other carnivores, but provide only about a third as much control over the deer population as is given by the single cougar on the same hunting area.

//The food and size-of-home needs of cougars and wolves are far more stable than the landscape. From the activities of these animals in the Rocky Mountain national parks today we can reconstruct the past, when many hills of early America each had a cougar and four wolves. Any one of those areas nine miles across would provide a fifty-acre home for each of 768 deer. From these deer, a third must have been "removed" annually to keep the herds in check and protect the vegetation for an indefinite future—the balanced state in which the white man found it. The cougar would have taken one of these deer every other day, the wolves one a week, leaving scarcely a fourth of the expendable deer for the assortment of lynxes, bobcats, coyotes, foxes, and occasional Indians. The pyramid of life was complete, and still allowed for the fact that the smaller predators depended largely on rabbits and hares, squirrels, mice, voles, and such birds as they could catch. This is the world man turned upside down. These are the deer he is now obligated to control.//

The outcome of protecting deer became evident soon after 1906 when Teddy Roosevelt, as President, created the Grand Canyon National Game Preserve in northern Arizona. His goal was the perpetuation of a particularly fine herd of mule deer and the magnificent plateau they occupied. Congress later incorporated into Grand Canyon National Park the southern fourth of the plateau, but the remainder is still national forest land.

The plateau itself is a natural deer park, an island of life

isolated from the rest of the country by mile-deep canyons (Grand Canyon to the south, Kanab Creek to the west, Marble Canyon to the east) and the arid, semi-desert lands of Utah on the north. For centuries it was a favorite hunting place and meeting center for Piute Indians and Navahos, and their legends referred to it as the Kaibab—"mountain lying down." They also called it "Buckskin Mountain" because of the ease with which they could obtain deer hides there.

The original beauty of this grassy and forested oasis is evident even in the solemn pages of the *Second Annual Report* of the U.S. Geological Survey (1882). In his account of its physical geology C. E. Dutton wrote: "There is a constant succession of parks and glades—dreamy avenues of grass and flowers winding between sylvan walls or spreading out into broad open meadows. From June until September there is a display of wild flowers quite beyond description."

From 1885 until the creation of the game preserve, stockmen with horses moved into the Kaibab, accompanying as many as thirty thousand sheep and twenty thousand cattle. Forage for the domestic animals seemed unlimited. But to protect the livestock, a systematic campaign against all cougars, wolves, bobcats, and coyotes was begun. By 1906 the plateau had experienced its first known overpopulation, through excessive herds of cattle, horses, and sheep. Man-sponsored competition threatened the deer.

Immediately after 1906 most of the domestic animals were moved out of the area and a still more vigorous program was introduced to make the Kaibab safe for deer. The major part of this work ended when World War I funneled off the professional hunters employed in killing predatory mammals, but the project was continued until 1931. By this time 816 cougars, 30 wolves, 7,388 coyotes, and 863 bobcats had been trapped or shot on the plateau.

Perhaps the preceding overuse of the game preserve by domestic livestock had reduced the forage for deer enough to limit their reproduction. In any case, with predatory mammals providing little competition, the deer herd increased by about a fifth each year, from approximately five thousand in 1906. Forest Service officials reported damage to the range by deer as early as 1918, when the herd had reached about forty thousand animals—one deer to each seventeen acres on the plateau's eleven thousand square miles. But at that time overpopulation of game animals had scarcely been considered. Too little was known about the habits and life history of deer, and of their relationships to other animals and to forage.

George Shiras III, the famous naturalist, visited the Kaibab in September of 1923 and was so appalled by the situation that he rushed a memorandum to the Forest Service: "Never before have I seen such deplorable conditions . . . but one conclusion could be reached, that from 30,000 to 40,000 deer were on the verge of starvation both in their summer and winter range." This letter is still in Forest Service files.

By the summer of 1924 the herd reached the enormous number of a hundred thousand—one deer to every seven acres. The Forest Service issued emergency hunting permits for the portion of the plateau under their jurisdiction, and license-holders removed about a thousand deer. The following winter an estimated sixty thousand deer died of starvation. Again the Forest Service issued permits, but in 1925 only about another thousand deer were taken. Once more the winter saw a sixty-per-cent loss to the herd by starvation.

For five successive years the permits handed out by the Forest Service allowed hunters to work on the herd. Yet the total they took barely exceeded five thousand animals. This seemed negligible by comparison with the winter kill of some ninety-six thousand deer in just two years. But the State of Arizona contested the right of the Forest Service to allow hunters to operate on the Kaibab plateau. The case went to the United States Supreme Court, where a favorable ruling encouraged the Forest Service to institute so-called "government killing." In 1928, 1,124 deer were removed in this program. In 1929 the number was 4,400, and in 1930, 5,033. By then the herd had shrunk to an estimated twenty thousand animals. This was still excessive because the range had been so extensively destroyed.

Since 1939, when careful control had brought the deer to about ten thousand animals—one for each sixty-eight acres—the Kaibab has recovered much of its original charm. Thousands of visitors annually come through the national-forest portion to experience the thrill of seeing Grand Canyon from its North Rim. Just as we have often done, they stand spellbound in the fragrant shade of great ponderosa pines, staring through the clear air toward the far wall of the immense gorge. Below them the canyon's eroded, color-banded slopes fall away, forming one of the barriers that keep the Kaibab deer on their plateau.

A few of these visitors realize that they are on historic soil. Here is the home of the group of large native wild mammals which first drew man's attention to the need for a balance between food to support a species and predators to control its reproduction. Many are still puzzled by the problems of administration. On the national-park portion of the plateau, the Park Service attempts to preserve natural conditions by exclusion of all domestic livestock and protection of every native mammal. In this fourth of the plateau, deer may not be hunted nor their predators disturbed. Beyond the park, in the Kaibab National Forest, the deer herd is managed for its scenic value, to keep it within the limitations imposed by the food supply. Deer numbers there are matched to the presence of some livestock under permit, and control is achieved through a program of annual hunting by licensed sportsmen.

As more and more people come to appreciate the need for man to take the place of predatory mammals in controlling deer populations, we might expect fewer conflicts over goals and jurisdictions like the controversy about the Kaibab. Actually, few have been resolved as smoothly as that over the herd Teddy Roosevelt perpetuated. Elsewhere a legislative tug-of-war is almost always going on, trying to meet the recommendations of people with divergent ideas about deer.

A small number in each state think of deer as a source of hides and venison. To the Indians on the Kaibab in former years, buckskins and doeskins were more important than the meat they could dry in the wind and sun. Indians still usually regard all wildlife as a resource to be utilized as needed. Beyond their reservations, where they must abide by the white man's hunting rules, their seeming inability to think as he does has led to astonishment on both sides. The manner in which a deer is killed appears particularly unimportant to an Indian, once its death has been authorized through a license.

Far more numerous are landowners who see deer chiefly as a menace to crops. Although they realize that the state will pay for demonstrable damage, they can feel mighty indignant over ravaged gardens, wrecked fields of corn, or serval years of forest seedlings trimmed to the ground, all by game animals that have a sort of diplomatic immunity. Deer can be frightened away or fenced out, but they must not be shot at. Usually, the state claims ownership of all deer, even if it accepts no responsibility for feeding them. The state is liable for the actions of its deer on private land, but not for the cost of deer-proof fences or the time required to prove a claim on crop-insurance money. Only in America is the landowner in this disadvantageous position.

More than four million men and 150,000 women in the United States hunted deer in 1955, carrying a license authorizing each to kill one deer according to rigid rules. Perhaps seven hundred thousand others hunted without any license and paid less attention to the rule book. Probably a majority of these hunters think of game animals as an excuse to tramp through the autumn woods and pit their trailsmanship against the wild target. To most of them, venison and hides are incidental. They would subscribe heartily to the words of the great outdoorsman Aldo Leopold: "The man who does not like to see, hunt, photograph, or otherwise outwit birds or animals is hardly normal."

The great majority of the country's population—nearly ninety per cent—fits into none of these groups. Most people would lift no hand against a deer. Instead, by their actions in crowding around exhibits of wild animals—even of local kinds—and in delightedly feeding "grown-up Bambis" in national parks, they show an aesthetic interest that is all too often overlooked. They would agree with another of Aldo Leopold's comments: "Babes do not tremble when they are shown a golf ball, but I should not like to own the boy whose hair does not lift his hat when he sees his first deer." As these people glory in watching a wild animal, they feel no urge to stuff the memory as a trophy on the wall. Hunters usually find this impossible to believe. And they can quite properly question whether non-hunters are shirking responsibility to the land enjoyed, or are merely being overly trustful that government officers will take care of any surplus wildlife. Yet the majority who wish to watch, admire, and photograph unhunted deer that trust man have rights that cannot be ignored.

The country's deer have to be controlled and, without enough native predatory mammals to accomplish this, man must assume the job in a respectable way. All wildlife populations need to be kept at levels compatible with man's uses of the fields and forests. Laws to unify these control measures are necessary, although it is difficult to have good ones if the public is apathetic. Moreover, deer can double their numbers in a single breeding season, changing the population faster than one can readily appreciate.

Today the predators seem to need protection far more than deer do. Many a naturalist suspects that predator-control programs are as obsolete as the draft horse. Most of the domestic animals the wild meat-eaters harassed are now kept indoors. Yet as recently as 1956, thirty-three of the forty-eight states had bounty laws for predators, at local or county or state level. Cougars are still being shot legally for sport on the Kaibab plateau.

Legislators often seem unaware that game laws become obsolete. In California today the hunters rarely take more than five to seven per cent of the deer herd during the annual shooting season. The law, however, chivalrously limits them to killing bucks. As in many other states, the prolific does are immune. And the unshot deer account for a net annual increase of between fifteen and twenty-three per cent each summer, only to have thousands die of starvation every winter. California still offers a bounty on cougars despite the admitted futility of the method and the fact that the state paid for deer damage in 1956 in almost every county.

State regulations in Iowa seemed reasonable in 1936 when the deer population totaled only about six hundred animals: no deer at all could be shot. By 1947 the herd clearly was on the increase, and by 1952 the estimate was 10,700 deer. It may be more than double that today. During the fall and winter of 1951–2 alone, so many deer crossed Iowa highways that 475 were struck down by automobiles. Every deer killed meant a wrecked car for which the state took no responsibility. Many people suffered serious injuries and others lost their lives. Still state regulations forbade deer-hunting. When thirty-eight more bucks and does were killed on the highways in November 1952, publicity stressed the unfortunate damage to the breeding stock rather than to the travelers and cars. Only a few people seem to have wondered in print whether this might be a heavy price to pay and whether the state might not now gain revenue by licensing deer-hunting.

When a state issues licenses for deer, the deer problems

are not automatically ended, although today the willing sportsmen can more easily arrange to go afield than they could a decade ago. The trouble is that more and more of each state's private property is being closed to all hunters because of irresponsible use of firearms. Many landowners are reluctant to make exceptions, even for fees ranging from twenty-five to a hundred dollars from each licensed hunter requesting permission to seek deer. The owners of the deer range would rather put up with deer damage which, when severe, can be charged to the state, than with hunter damage for which no rebate is possible.

Unharvested game and chronic deer damage often lead in the same direction. The outcome can be appreciated from surveys made in Jackson County, Wisconsin. There, in 1948, deer were endangering reproduction of many forest trees, and farmers found great difficulty protecting their crops in early spring. State conservationists admitted that the range was overpopulated, but relied upon the usual autumn hunting season to correct the situation.

In March 1949 the game wardens were sent out to make a census. Their tally was high—as usual. Yet they returned full of alarm over finding one frozen deer carcass for each 12.3 acres. Of these animals, nearly half had died of starvation, which was no surprise. A few gave no clues as to cause of death. But over forty per cent of the frozen animals— one deer for every thirty-one acres—had succumbed to bullets.

In the spring of 1957 the Wisconsin wardens made another check in the same county. While surveying one area from which licensed hunters had reported better than average luck, the wardens found over three thousand deer dead in the woods—ten fawns and seventeen adults "wasted" for every ten bucks obtained legally. Yet the herd, with all its does protected, had increased despite both licensed and illicit shooting. The wardens concluded that if waste were eliminated, two or three times as many legal hunters could get a deer each—still giving the same inadequate control over the herd.

Actually, the carcasses left in the woods are not *wasted*. They take the place of the carrion left by a cougar after each kill. A regular supply of meat from any source is important to the smaller carnivores that are so helpful to man in controlling rabbits, mice, and insects. Even a deer dying from starvation contributes to the balance of nature, whereas one carried off on a hunter's car is a true loss to the living community.

America seems to have plenty of hunters to control deer: nearly four and a half million with proper licenses, and a deer population of eight and a half million outside of national parks and wildlife refuges. Still this army of volunteers, who pay for the privilege, does not keep the deer in check. The country needs fewer deer, not more hunters, yet both imply still more disappointed purchasers of licenses and equipment. Perhaps it is time to wonder whether the laws and customs regulating American sportsmen should be changed to match the present imbalance between deer and natural predators.

Very different methods for handling surplus animals are in vogue throughout Europe. In Denmark, for example, game wardens reported 86,761 deer as the 1945 population. In the 1945 hunting season almost a quarter of these were bagged, with no limit for each hunter and a much longer season than is allowed here. The deer population remained constant, in balance with its food supply, as had been carefully planned.

The clue to this difference is that in Denmark, as in so much of Europe, all hunting rights are vested in the landowner. He and his guests may hunt legally; anyone else is a trespasser. Hunting is organized to remove the surplus animals and weaklings, and to regulate the proportion of the sexes as well as the age distribution. It is a rite with complex rules, not a right of anyone to whom the state sells a license. Hunting in Europe is almost entirely a prerogative of the rich, never a democratic pastime as in the United States and Canada.

Curiously enough, our forebears brought to America the hunting code of European aristocrats and applied it out of context. There it was a workable plan drawn up by wealthy gentlemen, permitting continuation or even improvement of small herds on private estates comparable to the George Reserve. Every owner had a few animals to harvest each hunting season, and he made of the event a social feast of considerable elegance. Deer were strictly a luxury, raised to be admired alive and then used in diverting sport and on the table. Never was a carcass wasted.

The colonists insisted on being gentlemen too. They would hunt in the same stately way. The deer, though abundant, should be treated as though America were a game park and the pioneers its owners. In Old England no gentleman would deign to lie prone to shoot a deer. This must not be done in New England either. No gentleman would fire at the glow of a deer's eyes reflecting a lantern at night; only a poacher would take such an unfair advantage of the animal. In Europe not enough deer were available to make such a technique thinkable among estate-owners and their guests. But the colonists decreed a stiff ordinance on that too, condemning a man caught jacklighting deer to as much as twenty lashes on the bare back, "well laid." So dire a punishment is no longer meted out, yet today's sportsmen in America still deplore any breach of the imported rules.

America's human population continues to grow, with needs for food and forest products increasing even faster because of a rising standard of living. Can our land supply these needs except through further curtailment of deer? A growth in the numbers of mankind implies more people with an urge to hunt. It implies also a majority who would enjoy watching deer, or thrill to the night call of a cougar and the soul-stirring chorus of timber wolves.

A yardstick is now at hand with which to measure deer, to relate them to their several means of control and to the land. It is not too late for man to use his yardstick in establishing a new living community, one in which he is a smoothly working part in balance with the other kinds of life. To do so is to behave as the intelligent mammal he claims to be.

Aaccording *to a Tlingit Indian legend, Porcupine (who controlled cold weather) was a great friend of Beaver, although neither of them could refrain from playing tricks on the other. Once Beaver swam with Porcupine on his back, and left him on an island. Porcupine had to freeze the lake before he could walk to the mainland again. Later Porcupine carried Beaver up a tall pine and left him there. The rough bark of pine trees today shows where Beaver's claws dug in while getting down.*

After Stith Thompson: *Tales of the North American Indians* (1929)

CHAPTER · 8

The Big Bark-eaters

❄

In the world of nature, as in a mystery story, small clues —easily overlooked—often yield the solution. "Tell me of what plant-birthday a man takes notice," wrote Aldo Leopold, "and I shall tell you a good deal about his vocation, his hobbies, his hay fever, and the general level of his ecological education." "Tell me how the beaver and the porcupine fare in America," another outdoorsman might add, "and I'll tell you how far man has progressed in rebuilding his world."

Today almost every state has at least a few beaver colonies. We take them for granted, without realizing how close beavers recently came to extinction. So often we overlook the declining numbers of a familiar animal until it is too late. Remarkably, the beaver has staged a comeback.//

At the turn of the century America was virtually without beavers. Only then did the states and provinces extend protection to this largest of the continent's rodents. Even this move almost failed for lack of beavers to protect. New York state had about fifty, Pennsylvania none. Supposedly the last beaver in Vermont had been slain with a hoe in 1800, and in the rest of New England only a few remained. The animals were so rare in Missouri and New Mexico and Oregon that, when reintroductions were considered, stock had to be brought from the sanctuary of Yellowstone National Park and from the wilder parts of Canada.

Fur-trappers fought the restrictions, although so few of the animals remained and the market for pelts was down. No other creature with so fine a fur could be caught with so little effort. Beavers walked and swam into traps as though unable to learn that peril was inherent in anything man had touched. And, in a sense, the trappers had history on their side. Soft, lustrous beaver pelts were the dark brown gold that led man to explore the temperate wildernesses of the New World. After New England's human population had reached a million and beavers there seemed extinct, an estimated sixty million of the animals waited to be caught farther west. To get these skins the trappers and traders worked even deeper into the back country, mapping watercourses as the first highways, long before settlers followed to claim and work the land.

The Hudson's Bay Company, chartered in 1670 by King Charles II, placed the beaver conspicuously on its coat of arms, and for more than two hundred years used beaver pelts as its unit of exchange. Indian trappers could tell at a glance how many "points" they needed to purchase a company blanket from the number of black bars of uniform length woven into the margins. These marks have become traditional in a "Hudson's Bay point blanket."

Today's Canadians encourage their children to emulate the industrious and resourceful beaver, and to honor the animal on the country's flag and coinage. For their land, intersected by rivers, the beaver was a logical choice to be the national emblem. Yet, as with the magnificent bald eagles in the United States, beavers in the Canadian wild were not spared and rapidly dwindled in numbers.

Beavers proved so trusting and easy to capture that it was easier for fur-trappers to eliminate them entirely from each area than to come into balance with a reduced population. Trappers actually helped prolong the supply of beavers by liquidating the wolverine—largest member of the weasel family—known also as the glutton and the "beaver-eater."

No doubt the Indians passed along to white men the knowledge that, almost single-handedly, wolverines kept beavers in check and were the real competitors reducing the number of pelts going to market. The wolverines accomplished this without diminishing their appetite for squirrels, marmots, nesting birds, ailing animals of large size, and carrion of all kinds. With a pounce and a crush of powerful jaws on the victim's throat, wolverines dispatched sick deer and unwary elk as well as moose, caribou, and bear slow enough to let the short-legged predator catch up. Small packs of wolves, healthy bears, and even cougars yielded their own kills to the beaver-eater and disdained to return for a later meal because the wolverine regularly defiled any remains with a malodorous urine.

Wolverines were credited with diabolical ingenuity in robbing traplines of bait and game without being caught. Often they sprang the traps and hid them too. Quickly the trappers developed a passionate hatred for beaver-eaters, an emotion scarcely softened by the discovery that wolverine fur is unique in repelling moisture. It does not become frost-covered when sewed around the face on the hood of a parka for arctic winter wear.

By the time legal protection was extended to the few remaining beavers, wolverines waited for them only in the most remote areas. Yet a new competitor had been introduced: domestic cattle. Along countless watercourses, cows grazed. Often they extended their diet of herbs and grasses by eating beginning saplings of aspen and poplar and the sprouts around old willows. Trees upon which beavers depended for winter food could not regenerate fast enough to get ahead. Moreover, dairy farmers resented having the pasture area reduced through the spread of ponds from beaver dams. Cows and beavers do not go together.

In the right place, beavers can contribute substantially to the welfare of the land. Seldom has this been plainer than in 1954, when drought turned eastern New York state into a miniature dust bowl. Right in the middle of the parched countryside Bear Mountain State Park remained green because of water conserved by more than twenty beaver colonies.

To a beaver, running water is something to be stopped with a dam. Working alone or in company, the animal hauls mud from the stream bottom, using this both as fill and as mortar. Interlacing sticks, some as much as eight feet long, serve as reinforcement. Soon the creek is modified into a staircase of ponds, the lowest and largest often eight feet deep when first formed. Later accumulations of silt make the water shallower and at the same time build up layers of rich loam. Over the millennia since the Ice Ages these repeated changes have done much to improve the country's agricultural soils. Indeed, an ecologist studying the decline of farming in Vermont has linked this to the state's lack of beavers. Without the aid of these natural conservators of water and builders of loam, the Green Mountains have lost their valley soil.

Beavers long antedated man in developing the split-level house, each a single-family dwelling. The famous chronicler Sir John Price wrote of these houses as "castells" [castles] at the time of Henry VIII, but in America we speak of them as beaver lodges. The several underwater entrances commonly open into a large, dome-shaped room used for dining and for drying fur. It is a few inches lower than the sleeping area, whose floor may be covered with grass and shredded wood. A broad doorway connects the two rooms.

The discoverer of the Yellowstone, John Colter, claimed to have escaped from a pursuing group of Blackfeet Indians in 1809 by diving naked into a creek and coming up inside a beaver lodge. According to his tale, he huddled there, shivering through the night, then hiked unarmed for two hundred miles—living on roots and berries—to reach civilization a week later. If his emergency shelter was of average size, he would have had a space four or five feet in diameter and three feet high, with smooth walls and with adequate ventilation provided at the peak, where the reinforcing branches are left unchinked.

All winter long the parent beavers and their young of the two preceding seasons live together in the lodge. The pond may be roofed with ice, the walls and ceiling of the house frozen into a cement-like mass far stronger than the mud and branches that have gone into its construction. Yet members of the family can emerge at any time of the day or night into the cold water flowing beneath the ice to reach a miniature forest of edible sticks and branches they have anchored through the weeks of autumn in the bottom of the pond. This is the community pantry, refrigerated but always accessible in times of hunger. From it each full-grown beaver will take daily about three pounds of bark, the equivalent of a young tree two inches in diameter at the base, but in lengths that can be pulled easily into the dining room of the lodge.

No one seems to have recorded seeing beavers mate in the wild. The event must occur in midwinter, for the period of gestation would be the same four months wherever the animals were, and the two to eight kits usually arrive in May. By then the two-year-olds in a wild family are reaching maturity, and the mother drives them from the lodge. Often the father leaves too for temporary quarters he has prepared in the stream bank.

In Europe, where a few beavers still survive after centuries of relentless persecution, no lodges are built. The animals are "river beavers," not "lodge beavers," and they occupy burrows in the bank like those used in America by expectant fathers. Does this imply that home life in the lodge becomes intolerable for him when births are imminent? They occur with only the year-old young of the previous season keeping the mother company.

The newborn beaver is engagingly alert, its eyes open, its body well covered with soft fur. It learns to swim from the lodge's feeding platform into the water of the entrance-way, and at six weeks of age is weaned. By midsummer it may weigh eight pounds. Adolescent beavers, at the time when they are thrust out of the parental lodge forever, have more than treble this weight, and the average oldster—aged between nine and twelve years—weighs between fifty and seventy pounds.

Most people recognize the work of beavers in a dam or a shaggy lodge in a pond. Outdoorsmen with keen eyes may identify beaver work as much as a thousand yards from the pond itself, in the form of narrow ditches radiating to food areas. These canals may have little dams that serve as locks, keeping the ditch full of water as an avenue along which the animals can tow suitable lengths of bark-covered wood cut nearby.

Beavers prefer the inner bark of poplar and aspen, but dine readily on birch, willow, and alder. To reach supplies high above the ground they will fell a tree four inches through in less than half an hour. Propped up on hind legs and flat tail, they chisel out great chips all around the trunk, then run for their lives as the narrowed bole gives way. Occasionally an animal is slow, or heads in the wrong direction, and loses the race. As a beaver can cut down a tree nearly three feet in diameter, the danger is very real.

Legend insists that beavers can drop a tree in a previously planned direction, usually toward a river or a clearing where the trunk will fall freely. They do show a preference for trees close to open spaces, and the trees usually topple as claimed. But the "plan" is a myth based on the fact that trees receiving light mainly from one side lean in that direction.

Few of the trees taken by beavers are commercial timber, and foresters seldom concern themselves over the presence of these animals. Strangely enough, lumbering operations or even occasional forest fires improve beaver habitat. Those trees whose bark appeals most to beavers are the pioneers that spring up and grow quickly until valuable kinds re-establish their dominance. So long as cattle are not pastured in the deforested area along a stream, the welfare of beavers is well served by these changes.

Beaver ponds invite many other wild animals. Deer and bear come down to drink. Moose wade to reach bottom vegetation growing in the shallows. Ducks dive and tip. Herons and bitterns stand at the edge and watch for frogs or fish. Turtles find a sloping log on which to sun themselves.

At the same time, moisture from the pond spreads into adjacent soils, sustaining the water table and favoring the growth of many plants attractive to land animals. Deer and elk benefit by the abundant browse around the pond margins. In the Colorado Rockies, wildlife manager Don J. Neff found far more food for these game mammals among the willows and aspens, in spite of the many meals taken by the dam-builders from these very trees. Along other streams, where old beaver ponds had drained through untended dams, only herbaceous vegetation grew, affording little to deer or elk. These relationships between browsers and beavers appear to hold true widely, and hunters of American big game should regard the beavers as their friends.

Trout fishermen in the mountains find that where beavers live, more trout can be caught. Each beaver pond is a carefully maintained home for trout, as well as for muskrats and an occasional mink. Not only does the pond form a trap for nourishing debris carried to it by the stream, but the rubble dropped by the beavers appears to enrich the water. In consequence, it supports a remarkable population of small creatures on which trout feed, over an expanded bottom where the fish can search for morsels. At high elevations the beaver pond traps the sun's warmth, raising the water temperature into ranges where the fish are more active and grow faster. Certainly more trout are found per mile of mountain stream in the calm waters just above beaver dams than in the series of smaller pools between one pond and the next. At lower elevations the converse may be true—the water temperature too high for trout, although waterfowl still thrive.

Beavers can overpopulate their streams just as readily as white-tailed deer on a game range. Like deer, they need predators or human control to keep their numbers in check. Now we are faced with the task of ensuring that beavers do not eat out their food supply.

With unlimited reproduction and winter food limited through excessive demand, beavers show habits undescribed in the accumulated lore of the wild. Sometimes the outcome is catastrophic for man, rather than for the beavers themselves. Remote reaches of an unnamed tributary of the Arkansas River in south-central Colorado provided a dramatic example. There in 1949 a pair of beavers was released beside the small creek, high up where they would have no competition from neighbors. By the next summer the farmers in the valley below were happy to learn of two dams constructed by the animals. Through 1955 the parent beavers and their grown young were maintaining a series of fine ponds, impounding substantial amounts of water. But without the controlling effect of wolverines or trappers, the beavers outstripped their food trees.

Through some quirk in beaver psychology, the animals did not remain in their valley and starve. Instead, the entire beaver population set out for the other side of the watershed. Every last adult and kit migrated for miles in search of satisfactory building sites. Behind them the dams fell into disrepair.

In the last week of April 1957 a dam far upstream gave way beneath the weight of water from the spring thaw. The torrent it released completed the destruction of the next barrier below. The combined deluge rushed down the mountain, carrying trees like battering rams, and took out every beaver dam all the way to the inhabited valley below. There a wall of muddy water rushed across roads and fields, engulfing human establishments without warning. The disastrous flood came during a cloudless period of early spring when no one had reason to be prepared. Some farmsteads were buried in silt to ground-floor level, and the creek scoured its own channel to bedrock.

To take care of legitimate complaints of beaver damage and to prevent recurrences of floods due to beavers abandoning their dams, temporary measures have been tried. State fish and game commissions have authorized limited trapping of "nuisance beavers," leaving the trapper free to choose his season and prepare the regulated number of pelts for gainful sale. In other areas, government agents have live-trapped the excess beavers for distribution by truck and airplane in remote valleys where exiles might be expected to thrive.

In relocating nuisance beavers, the mating habits of the animals must be allowed for. Ordinarily, each individual pairs for life and remains tolerant of close relatives, or even becomes socially dependent upon their nearness for continued occupation of a stream. If single individuals are transplanted, they rarely settle down in the new locality. The two members of a mated pair, moved together, are more likely to adopt a fresh site and make it home. If a few of their young are included with the mating pair in shipment, the wildlife officer can almost guarantee that the group will stay where man has planned.

Maintaining a whole beaver family healthy and ready for relocation is difficult, and the preference is to release its members on the same day that they are caught. This makes it easy to overlook parasites or even diseased individuals. An error may explain the die-off of beavers which began in 1953 throughout Michigan's "upper Peninsula." Although 323 dead beavers were examined in 1954, no one is sure what happened. In 1955 the disease disappeared, seemingly having run its course.

Concurrently, several trappers of beavers turned up with tularemia, probably contracted from the animals they had skinned. The bacterium of the dread disease was recovered from one beaver. This, in itself, was surprising, as beavers ordinarily are free of tularemia and of the ticks that carry the malady from one animal to another. But when beaver lodges in the Upper Peninsula were checked in 1955, many of them were found infested by the tick *Ixodes banksi*. Trappers knew about the ticks, yet could not recall seeing any on beavers of the Upper Peninsula prior to 1953. Were the ticks, and perhaps tularemia, brought in on relocated beavers?

Unless the wolverine and perhaps other predators are brought back, it seems likely that man will have to continue indefinitely his control of beaver populations. So long as the fur maintains a fair market value, this is easy. By 1933 six states found that beavers, almost extinct in 1900, had increased rapidly under protection. Short beaver-trapping seasons were tried the following year, and some 11,973 animals were caught. By 1951 thirty-one states had official programs for beaver, yielding 146,468 pelts in that year. In the same season seven of Canada's nine provinces supplied almost two hundred thousand beaver skins without damaging the breeding stock. Yet it is difficult to compare these figures reliably with the situation in colonial days, to see whether the beaver has come back to its original population.

Formerly, beavers were distributed widely from northern Florida to Newfoundland and the west coast, as far north as the limit of mixed forests. The British settlements along the Atlantic coast kept indifferent records, and the French who controlled the vague arc of land from eastern Canada to the mouth of the Mississippi were even less systematic. Only the Hudson's Bay Company far to the north kept careful tally, and in the years prior to 1877 showed sales of as many as a hundred thousand of these skins annually. Most of the pelts leaving American ports for Europe went into making beaver hats, the fashionable headgear of the day. These went out of style with the introduction of the high silk hat, giving beavers a reprieve just as they neared extinction.

Today the demand for beaver fur is almost exclusively for ladies' coats, but how long it will last cannot be guessed. Already the whole fur trade is seriously threatened by the variety of synthetic fabrics with luxurious pile. Many potential buyers prefer these to animal skins because of relative immunity to insect attack and also because of lower cost. Consequently the wandering fur-trapper of the romantic past is almost gone. With him goes much of the need to prevent any new spread of wolverines. Perhaps it will soon be time to encourage the beaver-eaters and have them take over their ancient role around the beaver ponds. In the last few years a number of outdoorsmen have praised wolverines and defended them against trappers' tales. Could this be a straw in the wind?

Most people hope the beaver has come back to stay. The nature-minded, camping public finds high entertainment in seeing beavers active at dusk and dawn. We ourselves have been surprised, at remote ponds, to find these animals taking up their old trusting ways of working around the clock, cutting trees and towing branches in full daylight as though man had never come. An occasional beaver, after an uneventful week or ten days of associating with us, has shown enough curiosity to swim over to where we sat and take a carrot or a piece of apple from an outstretched hand. How tame these individuals would have become after continued kindness is not hard to imagine. Few animals show such readiness to tolerate human neighbors.

It is ironical that, by bringing the wolverine as close to extinction as they could, the fur-trappers made possible the beaver's amazing recovery as soon as protection was given the few survivors. In less than fifty years beaver numbers have reached a point which probably exceeds that under wolverine control before the white man came to the New World.

The fur trade also changed the fortunes of another bark-eater, the waddling porcupine. But this second-largest of our rodents reproduces so slowly that nearly two centuries were required for it to become a menace. Porcupine populations today far outnumber any known from earlier times.

The animals endanger commercial forests and farm crops, as well as inquisitive livestock and overeager dogs.

An unwritten law of the outdoors has protected porcupines. They are the only large animals an unarmed, hungry man lost in the woods can kill with a stick. Without preparation, he can lay the carcass on a bed of coals, surround it with embers, and have a meaty dinner ready in less than three quarters of an hour. The flesh is white and tender, although a person must be really hungry to overlook the pine-bark flavor. In consequence, rumor has held that porcupines must not be killed.

No state or Canadian province actually offers the porcupine sanctuary beyond the boundaries of national parks and similar preserves for all kinds of wildlife. Today the increase of quill-bearers in protected woodlands is a growing problem. At Mesa Verde National Park in southwest Colorado the western yellow-haired porcupines are girdling many of the sturdy piñon pines along the mesa top and canyon walls, where these evergreens have long given the area a character. Oak brush (which is far from beautiful) and a nondescript juniper are taking the place of the pines, and park officials are much concerned. It is difficult for guides to impress visitors with the resourcefulness of the Indians who built Mesa Verde's cliff houses and create interest in former dietary use of piñon nuts unless piñon pines can be pointed to. If porcupines there are not controlled, this will become increasingly troublesome.

The eastern black-haired porcupines are even more destructive in the Quabbin Reservoir area set aside to protect Boston's metropolitan-district water supply. Dens with stratified droppings have been found on some of the islands in the reservoir, showing that large populations of porcupines once lived there. But after killing off the hemlock trees, the animals swam to the mainland and began attacking the plantations of evergreens there. Because little hemlock is available, they have turned to red pine and larch, with disastrous results.

Porcupines would be uncommon even in the deep forest if man had not removed almost all of the original controls on their numbers. Foremost among them was the medium-sized, weasel-like fisher, known also as the pekan. For more than seventy years the market for fisher pelts held at such a high level that trappers took between eight and nine thousand annually. They sought males weighing up to eighteen pounds, but concentrated on the smaller, six-pound females because their finer, softer skins brought as much as $150 each. The record payment for a single prime pelt seems to be $345, but even $150 was a lot for an easily trapped denizen of dense woodlands.

The fisher is peculiarly vulnerable to intense trapping of females, for its gestation period lasts about fifty weeks. Almost immediately after giving birth, the mother mates again and is pregnant while nursing her few young through much of the winter, the time when her pelt is most sought. Each one trapped represents a disproportionately large loss to the future of fishers: an adult, an unborn litter, and a little family that starves when the mother is caught.

In the United States, fishers are now among the rarest of native mammals. Yet only a few fish and game commissions have succeeded in completely protecting the survivors. Remote mountain areas, particularly close to the Canadian border, have a few. In those regions porcupines are even rarer, for fishers find them easy prey. Canada still has fishers and little trouble with porcupines. Now the market for fisher fur has collapsed so completely that since about 1940 the old-time persecution of these animals has ceased. Where any remain, they may repopulate the forests.

The lesser enemies of porcupines—wolves, cougars, bobcats, lynxes, and great-horned owls—have also disappeared from much of the continent. Without these checks, little families of porcupines gradually multiply, although the "quill pig" does not mature until it is two or three years old, usually raises a single young each year, and often dies of disease before it is five.

The baby porcupine is called a "porcupette" by Dr. Albert R. Shadle of the University of Buffalo, leading student of porcupines' early life history. It weighs a full pound at birth—several ounces more than the average beaver kit and twice as much as a black-bear cub. The porcupette opens its eyes immediately, and will raise its quills even before they are dry. Within a day or two it follows its mother out of the den and up whatever tree she chooses. To nurse, it must keep close to her and really insist on interrupting her feeding. Usually she tries to push the porcupette away, as though attempting to escape from further maternal duties. Within a few months she succeeds, weaning her offspring by leaving it to its own devices in the woods.

Thereafter each porcupine is a strict vegetarian, taking a wide variety of herbs and even aquatic plants during the summer as a supplement to the standard diet of buds, flowers, catkins, and bark. Where wood lots stand beside grainfields, crops of corn, or orchards, the animal can cause extensive damage. It is willing to sleep all day in a stone wall or a culvert, then amble half a mile for a nocturnal feast.

From the very beginning a porcupine makes distinctive noises. At first the mother communicates with her porcupette by clicking her teeth. When the baby is a few days old, its teeth are big enough to click in response. Soon it is making the grunting sounds for which the quill pig is named. (The word "porcupine" comes from the Old French *porc espin*, "spiny pig.") The porcupette's quills rattle like those of the adult as it ambles along. Even after a porcupine is on its own—and for most of their lives these animals are solitary—it continues to make sounds, as though talking to itself.

In mating season, male porcupines vocalize in a high falsetto, "singing" or serenading a female while following her around. Usually the prospective mate bristles at his every advance, and may squall so loudly that she can be heard half a mile away. But if he persists, her feminine indecision may be overcome. Then the pair will rub noses or use the forepaws in a fondling gesture, each porcupine caressing the other's face. A few hours after breeding, however, the female usually becomes antagonistic again, and for her any further amorous play is ended. Thirty weeks later her porcupette will be born, with only the mother in attendance.

When mature, a porcupine may measure thirty-six inches in length, with a six-inch tail. Among the long shaggy hairs on its back the animal will carry about thirty thousand air-filled quills, each held in place by a tiny bulb in the slate-gray skin. Every quill is perfectly smooth and glossy white except at the dangerous outer end. There the polished point is embossed by about a thousand tiny barbs that keep a victim from extracting the quill once it has been well inserted.

A vigorous swat of the tail may dislodge a few quills and toss them for several feet. This is the factual basis of the superstition that porcupines "shoot" their quills. Otherwise, the quills are attached so firmly that the animal can take advantage of them as floats while swimming. The fisher avoids the quills by reaching quickly under a porcupine and rolling it on its back, then biting in through the unprotected belly surface.

A person who knows how can capture a porcupine barehanded. The secret corresponding to Achilles' heel is a region of long hairs extending well beyond the quills at the end of the tail. If, when the tail is motionless, these hairs are seized suddenly in one hand and pulled firmly, the porcupine will make no attempt to back up. Nor will it try to turn and use on its captor the four chisel-shaped incisors like those of a beaver, although they would be quite capable of cutting a human finger in two at one bite. Instead, the porcupine will attempt to drag itself away in the opposite direction. The animal relies so completely upon its quills that no other means of defense is substituted.

We have found that, with the tail immobilized, the appendage can be raised easily and a hand reached below it, where there are no quills, to the quill-free base. There the porcupine can be held firmly from underneath, or even

lifted and carried by the base of the tail without special precautions.

Left to itself, a porcupine is never aggressive. Even when
molested, it tries to escape, or stands with quills exposed—
the cactus of the animal kingdom. Only when annoyed deliberately will the porcupine sidle toward an attacker and
strike with its quill-set tail. The latter can drive the quills
through a heavy glove or into a cow's nose so firmly that
removal is possible only with pliers. And each quill must be
tugged out separately. Without this first aid, the quills work
in deeper and deeper.

If any creature molesting a porcupine receives a large
number of quills in the face and throat, the act of eating
often becomes so painful that the animal starves to death
rather than endure the agony. Otherwise, the quills are remarkably inert. They are never poisonous, and rarely carry
an infection into the wound they create. However, they
won't come out of their own accord until they have passed
through an arm or a bulge of the body. They don't collapse.
Nor will solvents soften the embedded barbs.

Occasionally a porcupine becomes so excited that it
strikes its own quills. We watched one trying to scratch half
a dozen quills from its wrist, where each was well embedded
and already beginning to work through the muscles. This
reminded us again of Dr. Shadle, who accidentally received
a single quill driven deeply into his forearm. He let it stay,
and within less than two days was able to remove it easily
from the opposite side. It had passed completely through
his arm, and would soon have dropped free, leaving scarcely
a scar.

Porcupines become more sociable in autumn, and members of both sexes den up in sheltered crannies among boulders on rocky cliffs, from which they can travel every few
days for a big meal. Quill pigs have no need to hibernate, for
they can always reach a good meal of inner bark and buds

no matter how deep the snow gets. Often they make definite trails between the trees and the den, and show a preference for fast-growing, healthy trees of valuable kinds: hemlock and pine among the softwoods, sugar maple, beech, and yellow birch among the hardwoods.

With its sharp, chisel-like teeth the porcupine flakes off the outer bark to reach nourishment surrounding the sapwood. Great patches are gnawed away, leaving scars known as "catfaces." Or a large limb may be girdled completely. Medium-sized trees are often killed, one porcupine destroying as many as fifty in a winter.

A catface readily admits invading insects and fungi. And if the tree recovers, the broad scar prevents the downward transport of sugar from the leaves. The inner bark just above the catface may come to contain from twenty to three hundred times as much sugar as any other part of the tree. Porcupines act as though they know all about this. When they attack a tree with a catface on it, the subsequent meal is almost invariably right above the previous defacement, taking the sugar-filled cells. It is as though the porcupines deliberately put the trees to work for them and returned a year or two later to harvest the food stored above the unsightly scars.

At all times of the year porcupines are ready to gnaw on bones and discarded antlers, apparently to satisfy a need for calcium. They show also an insatiable craving for salty materials and will return repeatedly to chew on a greasy frying pan that yields this flavor. Summer cottagers frequently complain because porcupines chip pieces from ax handles, canoe paddles, even the steering wheels of open cars—anything touched by a sweaty human palm. Occasionally they are attracted to rubber and bite into the tires of cars parked in the wilderness. At least one motorist in New Mexico's mountain country had to return to town using his gears as brakes on all of the downgrades because porcupines had cut through the hydraulic fluid lines during the night.

The real economic damage by porcupines is to "tree farms," such as are being tried extensively by so many of the big lumber companies in the west. To combat this growing menace to the timber crop, foresters tried setting out salt blocks poisoned with strychnine. But the tannins from the diet of bark precipitate the poison, rendering it harmless. The porcupines of eastern states proved almost immune to strychnine, seemingly because their food contains more tannins than are taken by porcupines in the west.

Game managers in eastern states have been able to reduce the porcupine population by following the animals' tracks in the snow and leaving at the opening of the winter den a few apples dosed with sodium arsenite. The porcupines do not carry the fruit out of the den, so other creatures are not endangered. And the poisoned apples destroy every animal in the den within a few weeks.

Many of the big lumber companies of the west are trying to protect their tree farms by a combination of trapping, poisoning, and cruising snipers. In Wisconsin a more imaginative approach was tried in 1958. From remote regions where fishers are still at large, a breeding stock was brought in to start biological warfare against porcupines. At the same time nesting platforms were erected for great-horned owls. The owls occupied these sites readily and raised unusually large families, often catching young porcupines as meat with which to feed their fledglings. No one is sure how the owl manages. Perhaps it grasps the big rodent's unarmed head in one giant claw, driving the scimitar talons into vital parts. Whatever the technique, it proves successful.

New York and other states with large timber areas or farm lands and orchards menaced by porcupines are watching these experiments closely. It seems almost too good to be true that, by treating old-time enemies of porcupines as allies in his modern home-front war, man may wage a successful, inexpensive campaign. Success of fishers and great-horned owls would hold down porcupines without exterminating this unusual animal. Can these two manage alone, or must wolves and bobcats, coyotes and lynxes be brought back too if man is to relax in his personal feud with porcupines? If the Wisconsin tests work out, man will have made a good start toward restoring the original balance of nature.

I STARTED *in killing buffalo for the Union Pacific Railroad.
I had a wagon with four mules, one driver and two
butchers, all brave, well-armed men. . . . It was my custom
in those days to pick out a herd that seemed to have the
fattest cows and young heifers. I would then rush my horse
into them, picking out the fattest cows and shooting them
down, while my horse would be running alongside of them.
. . . I have killed from twenty-five to forty buffalo while
the herd was circling, and they would all be dropped very
close together; that is to say, in a space covering about five
acres. . . . I killed buffalo for the railroad company for
twelve months, and during that time the number I brought
into camp was kept account of, and at the end of that period
I had killed 4,280 buffalo.*

"BUFFALO BILL" CODY: *True Tales
of the Plains,* as quoted in
*America: A Library of Original
Sources* (1925)

‣✿‣✿‣✿‣✿‣✿‣✿‣✿‣✿‣✿‣✿‣✿‣✿‣✿‣✿‣✿‣✿‣✿‣✿‣✿‣

CHAPTER · 9

Keeping a Closer Count

❀

W HEN the twentieth century reached its midpoint, the
United Nations made a strenuous effort to learn the number
of human beings in the world and where they lived. Into its
Population Division and then into the Statistical Office
poured reports and estimates from member nations. Some
uncertainty remained when all the evidence had been ana-

lyzed. The earth might be supporting 2,471 millions of people, or the figure might be as low as 2,300 millions.

A round number of 2,400 millions seemed good enough for 1950. But what do these figures mean for other living things upon the globe? In 1950 the world supported also about 24,000 bison, 2,000 bald eagles, 500 trumpeter swans, less than 60 California condors, and just 32 whooping cranes.

It is not enough to know the total for mankind or bison. Twenty-four thousand bison—the animal Americans call "buffalo"—sounds like a goodly sample of the Old West, even when compared with fifty million bison three centuries back. The shock comes in discovering that only about four hundred of them are at large, or in learning that in 1906 not a single bison was known in the United States outside the ranch of Michell Pablo in the Flathead Valley of Montana. Our free bison then were extinct. Today's ten thousand plains bison in national parks and wildlife refuges are all descendants of Pablo's herd. An additional fourteen thousand roam Canada's Wood Buffalo Park, a wilderness half the size of Maine set aside along the boundary between Alberta and the Northwest Territories. They appear to be hybrids between the original small stock of the wood-bison race there and 6,673 plains bison transferred to the park between 1925 and 1929.

Change in numbers is far more meaningful. Rapid declines in certain kinds of animals in the 1950's led a prominent magazine to publish a picture spread on "Fossils of the Future," drawing attention to some of the verge of extinction. In the same decade people all over the world banded together into an International Union for the Protection of Nature, hoping to soften the competition between mankind and other forms of life.

The development of this human competition can now be traced with reasonable accuracy for about 140 centuries, including the three since America was opened up for settlement by Europeans. The United Nations statistics show astonishing changes during this most recent period. Since 1650 the Caucasoid peoples of the world—including the Arabs and the natives of India as well as Europeans—have increased from less than a fourth of the total of mankind to about fifty-seven per cent. The Negroid race has fallen, in spite of extensive transplantation out of Africa, to less than half the proportion known for 1650. The Mongoloid race has not kept up either.

Dramatic as these changes are, they cannot conceal one fact: the green plants of the world, which alone trap the energy of sunlight and use it in building foodstuffs useful to members of the animal kingdom, are supporting five times as many people today as in 1650. According to the famous anthropologist Dr. Margaret Mead, the number of people now alive is greater than the cumulative total of the world population prior to 1900. Living plants cannot be expected to support as many non-human animals as previously. The decline of the American bison and the extermination of the passenger pigeon are but tokens of this alteration.

Three centuries seems a brief period for such striking changes. Yet census figures do not show how much human habits have altered at the same time. Our U.S. Bureau of the Census has no space to record that we in Anglo-America, although a mere seven per cent of mankind, now consume fifty per cent of the world's non-renewable natural resources and a lion's share of the renewable ones as well. By doing so, the citizens of the United States and Canada unavoidably affect lives all around the globe.

Some of our most trivial-appearing habits have far-reaching effects. The way we buy our food and preserve it from spoilage has no place in the census figures. Nor is a record kept of the number of city dwellers who have moved to the suburbs or to larger apartments where they can easily keep a cat, a dog, or a parakeet. Yet these are all related. Pets have moved too, from a marginal status to a big business. No longer are they "eaters and snappers up of unconsidered trifles." The leftovers of "people food" are put in the refrigerator. "Pet food" is bought separately to feed animals indoors.

In America four homes in every ten now house at least one dog. Dog-owners spend more on dog food than is spent altogether on baby food. The makers of metal cans report that more of their products are filled, emptied, and discarded in the pet-food business than in canning any vegetable or other tinned product except oil and beer. Yet a large part of the food sold for dogs and cats is marketed dry, in bags and boxes. These facts reflect a rise in the dog population of the United States by at least thirty-five per cent in the last decade, to more than twenty-six million animals.

In Britain few homes lack a pet. Fifteen million homes now shelter a dog, a cat, a caged mouse, a hamster, a guinea pig, a rabbit, a fish, or a bird. More than $140,000,000 a year is spent on feeding them. The booklet *Advice on Pets*, issued at one shilling and three pence by the learned Zoological Society of London, has become almost a best-seller; it tells how to feed and care for everything from foxes to chameleons.

Most of our meat scraps and cereals for cat and dog food come from America itself. Suppliers find that pet-owners are eager to buy scientifically balanced diets, even at prices held high by competition with other users of the same raw materials. For millet and "canary seed," however, the pet-supplier goes to Africa. In our country millet is grown as hay. Its seeds are regarded as food for birds only where waterfowl crudely thresh the fallen heads. In Africa, by contrast, millet is a staple food for millions of dark-skinned people. They are willing to grow and collect millet seeds, raising large crops of it for sale to the bird-fanciers of the New World. "Canary seed" is the fruit of manna grass, another cereal of the Dark Continent. It grows well on dry uplands, and can be cultivated over vast areas if the market warrants.

The natives of southern Africa's high country are as eager for the possessions of civilization as anyone else. To feed the canaries, budgies, and tropical finches in British and American cages, they are willing to put more and more of their land into cash crops of millet and manna grass. They have, in fact, been led to raise these crops so single-mindedly that their own diets have changed. Instead of growing enough local vegetables or Indian corn for "mealies," they have turned to canned goods from everywhere and dry corn from America. Whether they can ever be happy again on their former diet is a question for the future.

As so often happens when man sets out to treat a field as a roofless factory, local animals take advantage of the sudden abundance of some few kinds of food. In Africa the millet fields and manna-grass patches have become mills catering to Caucasoid markets. In them the common African finches known as weaverbirds have found a bonanza. Suddenly, for them, the carrying capacity of the countryside has risen; never before did it afford so much bird food. The weaverbirds have multiplied and now raid the fields, endangering the cash crop upon which the natives depend for purchase of human food from the store.

Two and sometimes three crops of millet and canary seed can be raised in a single good year. But between the seasons of harvest the local finches must find food elsewhere. They have taken to raiding a wide variety of cultivated plants, many of them never attacked when weaverbirds were fewer. To protect these gardens and the crops of bird food, the human population now seeks out the roosts of finches to dynamite them during the night and bring the numbers down. Natives unaccustomed to so powerful a tool often get hurt. Some die. Yet they do so to defend their new way of life, their right to buy a sewing machine or a radio or a can of meat. They are risking themselves for civilization, and to save cage birds in America from starvation.

To many a naturalist it seems incredible that economy is served by raising Indian corn in America, defending it from European corn-borer and European earworm, then shipping it halfway around the world to feed African natives who could and would grow it themselves if they had to. During the past decade these inhabitants of the Dark Continent have become specialists, occupied with increasing the global supply of bird food, raising millet and manna grass that could just as well be maturing on American farms!

Intercontinental traffic of this kind is both expensive and risky. Every vessel arriving at our shores provides another chance for some pest or disease to complete a free ride, undetected. Malaria and yellow fever apparently reached the New World in this way, along with the mosquitoes that transfer them from one sufferer to the next; the insects supposedly reproduced in transit, living in the tanks of fresh water carried for human use.

Inspectors at our borders attempt to intercept these unwanted additions to the flora and fauna of America. Yet no one noticed the Chinese blight of American chestnut trees until 1904, when it turned up in New York state. Within four years it had spread to Massachusetts in one direction and to Virginia in the other. Soon it eliminated virtually every chestnut tree in the great Appalachian area. West of the mountains some chestnuts still grow to nut-producing age, but few indeed are the hundred-foot trees familiar to our grandparents. Usually each tree is doomed as soon as its bole grows large enough to attract the beak of a woodpecker. These birds carry spores of the disease fungus from infected branches to intact trees. No American chestnut seems resistant, nor has man discovered a control procedure.

The census-taker and the income-tax investigator can-

not look into such inconsequential details as possession of pets or a liking for decorative shrubs beside the house. Yet, while the human population of Bermuda fearfully watched the coasts for enemy submarines, some legally imported shrubs brought calamity to the islands. In 1944 a retired American industrialist who owned a handsome place in Paget Parish imported a few plants from a nursery in the United States. On them one or more pregnant females of a scale insect, *Carulaspis visci*, apparently rode past the local inspectors of the Department of Agriculture. Their offspring promptly spread to the islands' chief valuable tree, the endemic Bermuda cedar, *Juniperus bermudiana*.

At that time, as in 1612 when the first British colonists settled on the coral islands of Bermuda, the cedar formed handsome groves yielding a reddish-brown, knot-studded, aromatic wood. The evergreens also provided firewood, lumber, and an effective windbreak as well as privacy between the close-set houses. As many as two million cedars grew there, some of them reaching a height of fifty feet. So resistant were the fibers of cedar wood that thin cross-sections of large trees were used as flagstones, and planks of cedar were worked into fine, polished coffins as well as a multitude of souvenir items.

When the island entomologists were called to identify the insects killing cedars on and around the industrialist's estate, they found not only *Carulaspis visci* but also another American scale insect, *Lepidosaphes newsteadi*. The latter seemed potentially far more important, as it was a near relative of the oyster-shell scale, known to attack a great variety of plants, including apple, birch, lilac, privet, and willow. By 1945 the Bermuda legislature backed the campaign to exterminate both types of scale insect, appropriating £14,000 to foot the bill.

Scale insects reproduce for much of the year without need for mates. By virgin birth each female raises batch after batch of young under the protection of her shell, and the little ones are distributed by wind. They attach themselves, drill into a fresh twig, grow a protective scale, and repeat the process. To reach them under their shelters a really powerful insecticide is needed. But in Bermuda, chemical sprays harmful to human beings must be used with great restraint, as Bermudians rely for drinking water on rain collected in open catch basins. Fresh water is too precious to contaminate.

The Bermudians called upon their fellow Britishers in North America for help, and an experienced entomologist came from Canada with the best ideas available in the Insect Control Laboratory at Belleville, Ontario. He brought along some live ladybird beetles, which in Canada are effective foes of young scale insects and will hunt them out before the virgin-born offspring have had a chance to develop a protective cover for themselves. Immature ladybird beetles are just as voracious as their parents, and in the laboratory on Bermuda the use of these beetles promised complete success. Soon the Bermudian Department of Agriculture had nine people busy raising beetles to liberate where scale insects were most destructive.

Ontario has no lizards, whereas Bermuda is abundantly supplied with these insect-eating reptiles. The lizards stepped in and eliminated the ladybird beetles and their young as fast as the Department of Agriculture liberated them. No doubt the lizards, while stalking ladybeetles, often stood on the backs of scale insects that were sucking safely from a cedar branch. The cedars continued to expire.

The near relative of the oyster-shell scale died out mysteriously, but *Carulaspis visci* continued its deadly work. By 1949 it was apparent that no control was forthcoming, and that Bermuda's appearance would be better served by expending money to remove the unsightly gray trunks of dead cedars than by trying to protect the remaining live ones. First the House of Assembly voted funds to clear dead cedars off government-owned land. Three years later an additional allotment went for removal of the tree specters within fifty feet of all 180 miles of Bermuda's roads. By 1955 the swath was widened to one hundred feet on each side, and today the mournful task is about complete.

We set down at Bermuda's airport in early spring recently, without being warned that the colony's appearance had changed drastically since pre-war days. Fresh from the forested isles of the Caribbean, we stared aghast at the treeless landscape. The sun seemed brighter than ever, reflected from pastel houses. With nothing taller than a bush in sight, where now was the privacy, the quaint charm of the narrow crooked roads between stone walls?

Dead cedars are being used for fuel, and carpenters are

finding (albeit reluctantly) that many of the gaunt trees can be worked into good boards. But the evergreen shrubs being planted as replacements for Bermuda cedars are unlikely to supply the colony's needs for either lumber or firewood. Already both are being brought by ship to the islands.

Men whose livelihood seemed unrelated to cedars were affected by the change. Banana-growers, who used to do quite well, no longer have windbreaks to protect their plants. Wind shreds the leaves and dries them. Respectable crops of fruit cannot develop. Bananas too are imported now. Bermuda, in fact, has lost so much of the tropical atmosphere it offered before the war that its latitude is easy to remember. Due east of Charleston, South Carolina, these bits of coral stick out of the ocean close to the path of the warm Gulf Stream. A few infected shrubs, intended to beautify a paradise, led to the collapse of an illusion.

Landscaping an estate to improve its appearance has much in common with clothing a woman in luxurious garments. Both are expensive demonstrations of a higher standard of living. Yet the nurseryman and the furrier need not divulge where their materials come from. No central listing is kept. Only the furrier must know how the cost of a mink coat is divided among the suppliers of the pelts, the skilled labor used in making it, and profits for the designer and salesman.

Trappers who have offered to supply mink skins in return for a finished coat their wives would enjoy know some of the answers. About forty per cent is the best saving they can achieve, for the skins themselves are but a first step toward a fashionable garment. Even then the wife may be dissatisfied. Comparatively few mink today go from the trapline to the furrier. Instead, the skins come from mink farms, where captive animals of mutant strains yield better fur, in colors of greater appeal than wild mink afford.

Captive mink must be fed. They eat meat. A mink farmer

must provide a health-maintaining substitute for the fish and frogs, mice and young birds a wild mink would catch. His problem is to find a supply cheap enough to let him make a profit on the skins he sells. Slaughterhouse scraps are much too dear; they go in cans to America's cats and dogs.

In casting about for a substitute, the mink farmers followed the lead of pet-food makers and exploited a resource on western ranges. There a surprising number of mustangs —those small but hardy and half-wild horses—roamed at large. Some may have descended from horses the early Spaniards lost. Others came from lines the Apaches and other Indians rode when they chased bison long years ago. Whatever their origin, the mustangs found enough to eat. They belonged to no one.

Water for mustangs was often another story. We'll not forget the summer we took a back trail in New Mexico, seeking a waterfall of which we'd heard. Of course, the water fell over the cascade only after rains or for a few short weeks in spring. We found the place, but remember it for its stench. Indians with rifles had shot more than two dozen mustangs near the intermittent watercourse, to eliminate competition at pools for a scattering of cattle and one herd of sheep. The bloated bodies of the little horses littered the desert, turning an otherwise pleasant trip into an odorous nightmare.

Mink farmers stopped the waste of horse meat. They bought mustangs on the hoof at the lowest possible price, from anyone who would lasso them and drag them to a railway loading platform. Mustangs became diced horse meat, then ranch mink, finally fur coats for milady.

Mustangs also became harder to find. In the ten years following World War II an estimated hundred thousand of these wild horses and feral burros were captured and cut up for the pet-food and fur industries. Probably no more than twenty thousand remained, most of them in hilly retreats and on public lands. Mink-farm scouts hunted them out in light aircraft, often herding them toward waiting men in jeeps on flatter country by buzzing the animals and by other techniques of low flight. Humanitarians began to protest at the relentless cruelty with which the survivors were caught and handled on their way to the slaughterhouse. As a result, one after another of the western states enacted special legislation forbidding these practices on public lands. In 1959 the federal government followed suit with a wild-horse bill (Public Law 86–234) banning the use of aircraft or motor vehicles in hunting these animals or burros on federally controlled areas. Because the meat-hunters had been poisoning waterholes on public ranges as a means of driving the mustangs into the open, a further provision of the wild-horse bill banned this practice.

Still the demand for ranch mink continued. Could fishermen supply "trash fish" to take the place of horse meat? Unfortunately, the fishing industry was itself an ailing operation. Fishermen were suspecting that the shrinkage in the harvest of marketable fish might have been caused at least partly by the capture and sale of "trash fish" including the young of those valuable when mature. The fishermen advised the mink farmers to find a substitute for fish.

Ironically, the fur farmers turned next to men they had pretty much put out of business: the trappers. These outdoorsmen, they knew, tramped their traplines, skinning each catch on the spot and discarding the carcass. Could trappers be encouraged to salvage carcasses for use as food for ranch mink? But almost no trappers existed, except some who had turned to other occupations. Former trappers ran gas stations or chicken farms—anything to make a living. The market for furs had sunk too far to be profitable.

The mink-raisers sought to reward resumption of limited trapping. They learned that the pelt market would pay an average of seventy-three cents for the pelt of a white-tailed jack rabbit, and twenty-three cents for that of a black-tailed jack. By itself, this was insufficient to pay for the work of catching, skinning, stretching, and curing the pelt. But if the hunters and trappers could count on an extra ten cents a pound for the carcasses—fifty cents as a bonus for each five-pound jack—the offer might find some takers.

The lure of $1.23 for a white-tailed jack rabbit was so much greater than eight cents a pound for hen chickens that many a poultry-raiser closed his henhouse and took off with traps and rifle. Through several of our western states, jack rabbits bounded into first place as "Furbearer King" of the 1956–7 season. South Dakota alone yielded a million and a half white-tailed jacks and eighty-five hundred of the black-tailed.

Jack rabbits were subject to no census. But when hunted vigorously from jeeps, the hopping targets grew scarce. Soon even coyotes had difficulty catching enough to eat and, during the winter when insects were few, began attacking domestic livestock. Stockmen applied for state aid, asking for government hunters and trappers to eliminate the coyotes.

Even before jack rabbits diminished, the mink farmers needed more meat. They approached a company on San Francisco Bay. Could whales be caught? Surely whales belonged to nobody, and each afforded a mountain of muscle. Actually, while the Japanese and Americans and Europeans were busy with World War II, the whales of the whole world recovered somewhat from near-extinction. The California company applied for permission to take a few within easy reach of shore, and to cut them up as food for hungry mink on western ranches. Licenses were issued and the work began.

No one knows whether the supply of whales will outlast the popularity of mink coats. But once again a luxury possession has reached out to influence the welfare of many

forms of life which no one a century ago would have imagined related to it in the slightest way.

No adequate record is kept of the directions in which increased human populations and changed habits pull the great web of life. If the stockmen succeed with rifles, traps, and poisons in ridding the ranges of coyotes that formerly kept jack rabbits in check, what will the insects that would have been eaten by coyotes do to other crops man cherishes? Who will keep the rabbits in control when the fashion changes, ending the need for ranch-raised mink?

Only a few years are required to wipe a species from the earth. Nor can a Michell Pablo be counted on to defend a breeding stock from which another start is possible. The last passenger pigeon was defended too, but it died in a Cincinnati zoo on September 1, 1914, at five p.m.—without issue.

We have need for a continual census, both national and international, to keep us aware of even subtle changes in our living environment. The interplay of economics and swiftly varying goals of luxury-conscious people now affects the welfare of primitive tribes, wild animals, and plants all over the world. To keep a closer count might be easier on all of us.

124

> A T first we had the land and the white man had the Bible.
> Now we have the Bible and the white man has the land.
>
> Bantu saying

{+0+0+0+0+0+0+0+0+0+0+0+0+0+0+0+0+0+0+0+0+0+0+0+0+0+0+0+0+0+}

CHAPTER · 10

New Weapons

❃

WHEREVER we have been able to travel far from signs of civilization and mingle with native peoples, we have been impressed with how little real damage to wildlife and forests is possible before the introduction of steel tools. The ax and the machete cut back the trees. The rifle and the wire noose eliminate the edible animals. And now the wielders of these tools have a new arsenal as well: poisons and substances made lethally radioactive.

Even Henry David Thoreau, turning his back on civilization for a stay at Walden Pond, required steel tools. He borrowed an ax with which to cut pine trees for his house timbers and prepare firewood to keep him warm. He used spade and hoe to cultivate his beans, after paying a neighbor with a steel plow and team to prepare the land.

Thoreau's experiment in living is known all over the world, yet few understand its deliberate frugality. How many concur with the naturalist-author in his credo? "This curious world which we inhabit is more wonderful than it is convenient; more beautiful than it is useful; it is more to be admired than used." Lacking these beliefs, we tend to use civilization's weapons as Thoreau never did—heedlessly to exploit and destroy.

With disquietude we recall the fragrant smoke in some of Costa Rica's highest mountains, where illiterate poor people with ax and shovel made a mockery of a great national park—without disturbing its signs. Well-fed men in the capital city and foreign visitors had exclaimed in awe over the truly magnificent oak forest clothing the peaks. But one by one the trees came down, each to be cut into short lengths and buried in an earthen pit, to become charcoal and fragrant smoke. As soon as the Pan-American Highway made the parkland accessible, unmarked trucks began carrying off the charred remnants of the oaks in equally unmarked burlap bags. The charcoal-burners were paid a few colones. City folk had cheap fuel for their cookstoves. And Costa Rica's forest heritage melted away from the highway under a fog-like pall of smoke.

An ax and a shovel seem simple weapons with which to slay a forest. In the mountains of Middle America they suffice to bare the peaks and hurry erosion on its way. The animals the forest has sheltered disappear, and the soil shifts too fast for a replacing cover of vegetation to develop. After ruthless deforestation, the land often becomes a worthless desert.

In Anglo-America we point with pride to our redwood groves, our inviolate national parks and monuments and forests. But seldom do we recognize the privilege of these natural resources, or that their preservation is possible because we possess other resources in an opulence without counterpart on earth. In a nation where cooking is done with electricity or gas, the market for charcoal is too small to support tree-poaching. To turn a redwood into cash, a man needs expensive felling equipment, a giant truck, and a matching sawmill. He would have to be desperate to risk this capital investment for the sake of a few illegal trees. By comparison, what could a park warden or a court of law seize from Costa Rican hillmen wielding an ax and a shovel —cheap tools left for them by the truckers beside the highway?

Far more of our people understand the reason for national parks and have the means to visit and enjoy them. Rarely does an Anglo-American see the law protecting trees as a barrier to bare subsistence for his family. Nor can education alone suffice to make a man respect public property. His stomach must be full too. And over much of the world it will be difficult (if not impossible) to raise both the standard of living and appreciation for resources before the native plants and animals are just memories.

In our own country it has been hard to convince gunners on Cape May, New Jersey, that the hawks which provide such unpredictable, darting attractions in autumn skies have not merely been hiding in nearby woods throughout the summer. To some of these men, hawks are vermin. To others, each bird simply affords a challenge to marksmanship. All of these people were incredulous when told that Cape May is an exception, a funnel for migrant hawks from the eastern states. Local abundance, they assumed, was merely part of general abundance. Yet until the practice was stopped, a single day's shooting often eliminated as many hawks as live in the entire state of Maine.

Our country is rich in people enough like Thoreau to insist on having wildlife and forests left to be admired, not used. They band together in the National Audubon Society. Some of them find employment as forest rangers, fire lookouts, and park naturalists. Our newspapers and magazines keep us informed on conservation efforts, and on changes in the numbers of animals and plants. Sometimes we credit this information service with being universal. Yet only a fraction of the Caucasoid race is even interested. The majority of men have not yet reached this plane, although they may claim to be civilized and pride themselves on the use of neckties, cars, and guns.

The weapons of civilization, whether ax or A-bomb, can be taken to the frontiers far faster than the point of view which will support a national park or wildlife preserve. Natural resources are imperiled when these weapons come into the hands of the general public. So long as they can be afforded only by the wealthy few, the damage may be slight.

Our migratory birds experience these differences as they wing southward in autumn. Through Canada and the United States, hunting pressure on waterfowl is regulated by flyway systems under a master plan that will allow an adequate breeding stock to make the round trip. Songbirds and most of the shore birds are forbidden targets until they reach Latin America or the West Indies. There they fly from under the protection of the migratory-bird treaties and meet new hazards. Poor people build ingenious traps and use sticky birdlime (made from the juice of local plants) to capture songbirds for the soup pot. Rich people entice the migrants with live decoys in artificial ponds and blaze away with shotguns. A single hunter may bag as many as two hundred ducks on a Sunday afternoon. It makes no difference whether he speaks Spanish in Honduras or English on Barbados. The only salvation of the birds is that men with guns and ammunition are few.

As soon as firearms come into the hands of the poor, wildlife is in trouble. Already in Ghana and many other parts of Africa the edible birds and mammals are gone. Even a cane rat may bring better than two dollars from meat-hungry natives with more cash than there are available proteins to spend it on.

The great migratory herds of caribou—between one and

two million animals in arctic America—began to vanish as soon as rifles were sold or lent to Eskimos. Until that time, barely over a century ago, the herds had moved each summer into the coastal tundras and returned for winter to taiga country where lichens throve, sheltered somewhat by evergreen pyramids of spruce and balsam fir. The Eskimos, waiting, watchful as the herds passed by, took enough of the big animals to supply modest needs for hides, sinews, meat, and bones. Yet with only homemade weapons the hunters had far less effect than the wolves in controlling caribou populations.

Eskimos with caribou were as independent as the Plains Indians had been with bison. Eskimos were reluctant to leave their families and trap fur-bearing animals for the white traders so long as caribou trotted past their camp sites in spring and fall. But what Eskimo could resist the chance to use a white man's rifle to kill more caribou? What Eskimo could foresee the decimation of the herds and his growing dependence upon selling furs to the cunning traders? Before the white men came, there had been bad years when few caribou arrived. Surely the animals had merely taken a different route and some other village had prospered.

At the same time, gold-miners and white trappers spread over the Arctic. Because of their presence, man's oldest weapon—fire—ranged uncontrolled among the taiga trees. Eighty per cent of Alaska's original white-spruce forest burned, and with it went the lichen carpet upon which caribou depended for winter food. Tundra—the summer range —will not burn; it is too wet and frozen. Yet, no matter how lush the tundra grows in summer, the caribou are doomed unless they can find lichens by pawing through the winter snow. Winter is the critical time for arctic animals, and "caribou" means "shoveler."

Off arctic coasts the hungry Eskimos helped white men hunt whales and walrus until these meat resources shrank

away. Sometimes the whalers stayed through the winter and sustained themselves by shooting caribou and musk oxen. Thousands of Eskimos starved, or died of white man's diseases.

With improvements in communication, the plight of Alaska's Eskimos became evident to responsible people. Still the underlying causes remained obscure. No one realized that the caribou herds had been decimated by the newly equipped riflemen, or that the surviving animals were starving because extensive fires had destroyed their winter food. The lowly lichen was not appreciated, and how seriously arctic weather limits its growth was not understood. From fifty to three hundred years are needed to replace a winter range after it has been destroyed.

To rescue the Eskimos, the U.S. government imported

reindeer from Lapland, and Lapp herders to teach the Eskimos how to handle these domesticated European caribou. The Eskimos were expected to change from opportunistic hunters to observant herders. For more than two decades after 1890 the importation continued despite warnings from the Lapp herders, who saw that little winter range remained to feed either caribou or reindeer, and that the Eskimos declined to follow a nomadic life, driving the herds on as soon as the lichens showed signs of depletion. Reindeer lack the migratory instincts of their wild forebears, and will not travel of their own accord even if they destroy the lichen carpet as completely as any fire could.

A total of 1,280 reindeer were imported before 1903. In many areas the animals prospered. As each herd grew, it was divided and driven along the arctic coast to become a meat-producing gift for another village of Eskimos. As long as this sharing with new regions continued, all went well. The Lapp herders returned home. But the Eskimos quickly forgot a warning: never let the herds become too large. By 1932 Alaska had nearly 650,000 reindeer. They ate out their winter ranges and starved. Their numbers crashed. Today less than five per cent of the peak population remains in Alaska. Moreover, this man-brought competition had served as a weapon, causing American caribou to dwindle even more.

In many places the destruction of lichens by fire, reindeer, and caribou has changed the taiga forest floor still more radically. Annual grasses have moved in, providing a rich pasture for summer grazing but contributing nothing the caribou can use when snow is on the ground. The lichens may never come back in these areas; summer range has spread at the expense of critical winter feeding areas.

On St. Paul Island in the Pribilof chain stretching westward from Alaska the U.S. government introduced four bucks and twenty-one doe reindeer in the fall of 1911. The starving native residents were asked to keep a close account of the herd and warned never to let it grow. They took a census every year, but killed only animals actually required for food. The herd prospered: 472 reindeer in 1931, 1,185 in 1935, and 2,046 in 1938. No one seems to have recognized that this was the peak, or that the herd was starving, having eaten out its lichen range. In 1940 the survivors numbered 1,227, in 1946 only 240, and by 1950 a mere eight animals.

A reindeer needs about thirty-three acres of average arctic range for year-long use, and the population on St. Paul reached three times what the island could support. The lichens left after the reindeer population crashed were insufficient to feed any herd from which the islanders could get the meat they needed. By disregarding the food chain linking a lichen supply to their own welfare, the people let reindeer become a weapon against mankind.

On the opposite side of the world the natives of the Dark Continent have leaped in too few years from the Stone Age into an uncomprehended form of civilization. Many of them live in the last natural big-game country on earth. Yet prior to World War II the natives benefited little from the presence of these spectacular animals. Only the graceful Thompson's gazelle was delicate enough to be held by a noose made of plant fibers, and the success of pit traps was limited. This, however, was no measure of intent. Meat is most attractive to the African, and wild game signifies meat. The Swahili word for the one sounds like the word for the other. Only the difficulty the natives encountered in taking game restricted their dietary proteins so largely to those from vegetables, eggs, and chickens.

During the war many natives went into military service and came to enjoy white man's food, with plenty of meat. When they returned to their villages, these ex-soldiers found their former diet inadequate and sought to bolster it either by legal hunting or through poaching in the game preserves.

At the same time, the veterans found farm land and pasture extending far beyond anything they recalled from 1939 or 1940. Wild animals were compressed into compact bands and were far more vulnerable.

Even the people remaining at home in Africa during the war received an education on food. Italian prisoners interned in the colonies soon longed for more meat than they were given, and taught natives to snare and trap game for the table. They showed Africans how economic it was to spend from cotton earnings a mere thirty shillings ($4.20) on ten lengths of light-gauge wire, fashion these into snares, and trap meat worth three shillings per pound dry. A single catch paid for the wire twice over. Profits from a single season of spare time could be enormous. Prison sentences meted out for poaching from game preserves were mild enough to be inconsequential, especially in view of the small risk of being caught in remote areas.

Ten shillings would buy enough extract from the bark of the "Hottentot's poison bush," *Acokanthera venata*, to coat twenty arrows. As investigators from the Kenya Wild Life Society discovered, this technique could be astonishingly inexpensive, for a minute quantity of poison entering the victim's blood stream causes death, and arrowheads can be dug out for reuse without additional coating. Ten shillings' worth should last the most persistent hunter for at least a year. To be successful, the bowman had only to conceal himself at a water hole and pump arrows into a herd of game animals as they came to drink. A skilled hunter could release from ten to twenty arrows before the herd stampeded out of range, and then follow the victims at his leisure. Circling vultures would guide him to the farthest dead animals.

Conservationists interested in the survival of Africa's spectacular wildlife have been especially distressed about hunting by men who do not use the meat. Arrow poison affects the edibility of the animal only at the point of contact. But often the hunter's sole object in killing wildebeests, for example, is to obtain their tails for sale as fly-swatters.

The extent to which these new techniques were adopted in Africa was revealed by the custodians of Tsavo National Park, Kenya's principal game sanctuary. Originally, this reservation of 8,069 square miles was one of Africa's finest for elephants. An estimated three thousand made their home within the park's boundaries. In each of the years immediately preceding 1956 at least six hundred of these animals were killed by poachers. At this rate the remainder would soon have disappeared.

Through its two thousand members the Kenya Wild Life Society applied pressure to the administration of British East Africa to eliminate the poachers from Tsavo Park. By August 1957 a total of 429 persons had been convicted of offenses against the game laws and the gangs were believed to have been broken up. An abundance of steel-wire cable snares and bows and quivers of poisoned arrows was found and confiscated. Hundreds of stretched skins, thousands of pounds of dried meat, and literally tons of ivory were collected from poachers' hideouts. These represented great numbers of buffalo, duiker, eland, giraffe, hartebeest, impala, klipspringer, reedbuck, roan antelope, Thompson's gazelle, topi, wildebeest (gnu), zebra, lion, and ostrich. The ivory from elephants proved that six hundred annual losses to poachers was a gross underestimate. When the yearly destruction was recalculated on the basis of the evidence and all of the different kinds of victims were totaled, the figure came to a minimum of 150,000 head. Had this practice not been curtailed sharply, the whole game population of Kenya would have come to an end.

Much of the ivory from poaching passed surreptitiously from hand to hand, at little expense and little gain, until it entered commerce along with license-covered tusks at

Mombasa, Kenya's port on the Indian Ocean. How many tourists buying ivory carvings at bargain prices wonder about the source of the ivory? How many African governments, once granted independence, will protect their heritage of wildlife and prevent a recurrence of systematic killing?

It would be easier to criticize the African natives' lack of foresight were we not guilty of similar actions on an even grander scale in America. For every tribe of black men crouching with poisoned arrows at a water hole, we have a white man piloting an airplane, spraying the landscape with deadly chemicals. The African kills only a few game animals. The American destroys mammals, birds, amphibians, fish, crustaceans, spiders, and a host of useful insects, while aiming at a few insect or fungus pests.

> ... *many of the so-called comforts of life, are not only*
> *not indispensable, but positive hindrances to the elevation of*
> *mankind.*
>
> HENRY DAVID THOREAU: *Walden* (1854)

+O(O(O(O(O(O(O(O(O(O(O(O(O(O(O(O(O(O(O(O(O(O(O(O

CHAPTER · 11

Herdsmen's Pride

ON our first trip to the American southwest we jour-
neyed from the forested piedmont of Virginia, through
Texas to New Mexico, Arizona, and Utah. Progressively
the land grew more parched. Trees gave way to desert
shrubs and cacti. We felt very far from home.

At frequent intervals in the arid west the road below us
buzzed briefly as the automobile tires crossed a cattle bar,
passing from one immense pasture into another. Occasionally a herd of red-coated, white-faced animals came into
view. Oftener we traveled for miles without seeing a cow, a
person, a house.

Cattle country, we learned, included essentially all areas
whose vegetation would support one or more cows per
section. A section, 640 acres, is a square mile. Land too poor
to support one cow per section was sheep country or desert,
depending upon who looked at it.

In New Mexico, Arizona, and parts of southern Utah
twenty-four thousand square miles of sheep country is

home to about eighty-five thousand Navaho Indians. This arid land was set aside in 1868 as a reservation for what has become the largest Indian tribe in the United States. The area was selected to accommodate the Navaho population at its former smaller size, and to correspond to the preferences of these Indians for a semi-nomadic life as successful sheep-herders. Reservation land seemed suitable for a sheep-based economy, with food and space for more animals than the Navahos could use. The Indians should have prospered.

Instead, they changed their ways and became sedentary. For a home base and winter quarters they built earth-covered one-room houses and called them *goghans*. The white settlers corrupted this pronunciation to *hogans* and overlooked the altered habits of the Indians. Or they rejoiced that the Navahos had ceased wandering and that each could be found at some fixed address.

On sheep country, sheep must never settle down, winter or summer. Their ability to thrive on land too poor for cattle comes from the way they use their teeth. They nip the vegetation and can cut it cleanly almost to the level of the soil. To maintain itself, the plant must then call upon resources below ground and produce a new crop of leaves. If a sheep eats these too, the plant has no chance to carry on photosynthesis and to store food. Soon it may die. All plants in sheep country die if the sheep are not driven onward. This is the basis of the sheep-made deserts of the Near East, of Spain, of Mexico, and of the Navaho land. Without plant cover and the network of unseen roots below ground, the soil erodes to gravel and bare rock on which nothing useful to sheep or man will grow.

In pioneer days the Navahos earned the ire of white settlers by collecting sheep—anybody's sheep. They still collect sheep—their own now. Sheep have become a prestige possession, a measure of an Indian's worth. A flock of thirty may provide all of the meat, hides, and wool a family of Navahos can use. Yet by husbandry and purchase they build up the flock to a hundred, two hundred, three hundred. The wool is sold to buy more sheep.

Sheep are capital, but not like money in the bank. Money strengthens a savings institution, although it may burn a hole in a trouser pocket. Sheep, beyond the carrying capacity of the land, destroy the soil. The distinguished Texan chronicler of the west, Dr. J. Frank Dobie, could think of no place on earth that offered "a more sterile-appearing sight than some dry-land pastures of America with nothing but sheep trails across their grassless grounds."

When bison or caribou graze an area until the vegetation offers little more, they move on. Sheep do not. Their wild instincts have been bred out of them. Man has made them "silly" sheep to save himself trouble. At the same time he has converted these meek creatures into a form of life far more dangerous than any lion. They may never attack any other kind of animal. Yet they inherit the earth by making it worthless.

Sheep destroy the cover a quail must have. They eliminate the grasses a mouse or prairie dog requires. One by one they eradicate their animal neighbors—even the predators that occasionally prey on them. Without quail and rabbits, mice and prairie dogs, no coyote or wildcat or eagle will remain long near a flock of sheep. The badgers and foxes go elsewhere. The skunks and ringtail cats move out. Soon only sheep and their sedentary shepherds are left upon the barren land.

The Navahos are proud of their sheep, but they cannot be proud of their reservation. No longer is it the rolling, grass-clad country they were given a century ago. Erosion has cut away the bared soil in many parts, turning the area into a national disgrace. Other parts go completely unused. Mostly, these are remote from roads and have scarcely been

seen by any man on the ground since the Navahos ceased
their wandering.

Soil conservationists from our Department of Agriculture have shown the Indians the relationship between overgrazing and soil destruction, and tried desperately to promote nomadic herding within the reservation and use of its
hinterlands as they were intended to be used. Advisers have
come to the Navahos from the Department of Interior's
Bureau of Indian Affairs. These men have pointed out that
no area in the west can support the large population of
sheep the herdsmen keep together, and that the economic
needs of all Indians on the reservation would be well served
by a fraction of the number. Still the Navahos cling to
their flocks. Their sheep remain a prestige possession, destroying the land and the other animal life.

To the eyes of easterners like ourselves, the eroding hills visible in Navaho country have a wild beauty, an other-worldliness in reds and browns and yellows. Evening light may lend purple to the landscape—every color except green. These are the hues of a New England autumn, just before the trees stand bare and gaunt as though dead. But the misused sheep country *is* dead in many parts. Centuries of care would be required to bring this once fertile region back to production of the many plants and animals an arid area can support.

The Bureau of Indian Affairs sees another aspect of Navaho methods, one often overlooked by soil conservationists and the annual visitors. Navahos are now recognized as responsible for increasing dietary deficiencies among the Hopi Indians. The four thousand Hopis live on rocky mesas, surrounded by Navahos. For centuries they have perched there, descending to the encircling plains only for water and to carry on a highly skilled dry-land agriculture. Hopis have always supplemented these plant crops with the meat of wild animals caught nearby. In this way they achieved a balanced diet. And until the present century the Hopis managed very well indeed. Few Indian tribes could feel so independent.

Before the white man insisted on law and order in the west, the Navahos wandered widely as hunters and food-gatherers. Periodically they raided Hopi fields. This stopped when the Navahos settled down and began to trade with the Hopis, offering woolen blankets and rugs in exchange for corn and beans and pots. The Hopis had little difficulty keeping Navaho sheep out of their gardens. But deteriorative changes beyond their fields were something else again.

These changes are so great that they might be charged to the distant past rather than to recent years. Yet records show that "the present desolate shifting-sand area that lies between the Hopi villages and the Colorado River was such good pasture land late in the eighteenth century that Father Escalante, returning from his canyon exploration, rested his travel-worn animals there to regain flesh."

The pastoral practices of the Navahos and of some white men on the plains around the Hopi villages have pushed the plant succession toward desert and greatly reduced the game-carrying capacity of the soil. Now the Hopis can catch very little fresh meat, and rely more upon their plant crops. They suffer from a shortage of protein in their diet, and an excess of carbohydrates. Even though their stomachs may be full and their bodies rounded, starvation of an insidious type is whittling down their efficiency, their ingenuity and resourcefulness—traits essential to life on the fringe of desert. Without raids or warlike acts, the Navahos and their prestige possessions may soon expel the Hopis from their mesa strongholds.

In these days of rapid travel by air, a person can visit the Navahos one week and the Masai tribe in Africa the next. Although almost half a world away, he can recognize the same situation in a new guise. The tall, dark Masai collect cattle, and sheep as well. They pasture them on the great savannas of Kenya, Tanganyika, and Uganda. The dire effect is most obvious to a person who returns to the Ngorongoro Crater near the vast Serengeti Plains between Nairobi and Lake Victoria.

Before World War II the Serengeti was a photographer's paradise, a plain dotted with herds of antelope, giraffe, and zebra. For the benefit of tourists, an attractive rest camp waited on the cool rim of the great crater, in the evergreen shade of giant *Podocarpus* trees. From this vantage point the visitor could look down to Ngorongoro's floor half a mile below, where thousands upon thousands of game animals grazed. The knee-high grass afforded the right amount of cover for lions, and a pride of as many as twenty might share in a single kill.

To protect this wealth of animal life for all time, the British administration set aside a vast area as national park land in 1940, and enlarged it in 1951. Within the park the game would be free to roam unhunted. Already the tourist trade out of Nairobi had become the colony's third most important industry. With encouragement extended to both visitors and wildlife, surely the whole region would prosper.

The British saw no reason for the new park system to interfere with the lives of African people in these areas. After all, the human and game populations had lived side by side for untold generations. Into the Tanganyika National Parks Ordinance of 1951 went the clause: "nothing in this Ordinance contained shall affect . . . the rights of any person in or over land acquired before the commencement of this Ordinance."

No one seemed to realize that between 1933, when a London convention had urged the formation of these park areas, and 1951, when the dream came true, the Masai tribe had grown amazingly. Within the Serengeti Park at the time of its reconstitution some 7,800 tribesmen were living together with approximately 115,000 cattle, 200,000 sheep and goats, and 6,000 donkeys. These people were following the age-old custom in Africa—owning livestock personally but pasturing them on community land.

The British have a penchant for making rules and abiding by them. But no one had made a list of Masai men entitled by the ordinance to stay in the park. And when the years following 1951 proved to be dry ones, no count could be kept of the additional Masai who migrated with their animals into the park from outside its boundaries.

Today it is easy to reconstruct the events leading to the impasse. The administration had done its best to improve conditions for human inhabitants of British East Africa. Tribal warfare was stopped, ending the former loss of life and destruction of property. Modern medicine helped the natives live better and longer, with a stability they had never known before. Their children survived in far larger numbers than they had under the care of witch doctors alone. And the diseases of domestic livestock were eradicated or brought under control with special success.

Probably the Masai profited more than most, for these lean, able-bodied people depend almost exclusively upon a diet of blood, milk, and milk products derived from their cattle. Masai herds multiplied impressively and the herdsmen felt rich. In this area of Africa, cattle form a medium of exchange, a mark of wealth. A man buys his wives from their fathers at so many cattle for a girl. To own a large herd of animals is like having a big bank account. The bulk of Masai cattle, in fact, provide only a resource, a mark of prestige. No product from these excess animals is consumed by man.

By striving for quantity rather than quality, the Masai have increased their herds only to lose out. Other tribesmen have done the same. During the past fifty years the number of cattle owned by Africans in Kenya, Tanganyika, and Uganda has increased about eightfold. The earliest reliable census, dated 1909, showed that the value of the cattle then averaged about £12 ($60) apiece. Twelve pounds was a lot of money in those days. But by 1957, when the latest census was completed, the total valuation of eight times as many animals was seen to be almost identical with that of forty-eight years earlier. Due to depleted ranges, the quality of individual animals had fallen until each was worth only an eighth as much—about thirty shillings. At the same time, the worth of the pound sterling had fallen; thirty shillings ($4.20) is very little money in these days. Inflation has hit the Masai.

Like sheep and unlike money, the cattle demand nourishment and water continually. To supply these needs, the

Masai anxiously drive their charges from place to place in search of grass and water holes. Eight times as many cattle plus a similarly augmented number of sheep and goats require vastly more food and water. The Masai claim these amenities for their herds in the Serengeti National Park under the terms of the 1951 ordinance. Bitterly the tribesmen criticize the park administration for trying to save the spectacular wildlife for which the lands were actually set aside. What good are game animals in the park, as no one may kill and eat them? Why should an African see his cattle starve or die of thirst merely so that white men can come and stare at wildlife in the reserve?

In drinking habits alone, the domestic animals differ greatly from the wild kinds. Cattle and sheep and goats must make repeated trips to water holes, and can never range far from these centers. In parading back and forth to drink, they destroy large amounts of valuable forage. Antelopes and zebras, giraffes and elephants, by contrast, manage for a week or more between visits to a water hole. Coming by less-frequented trails, they do relatively little damage to vegetation along the way.

Domestic animals consume far more water per pound than any of the African game, and this inefficiency works to the disadvantage of wild creatures. Every cow displaces nearly twice its weight in native wildlife. And consistently the Masai drive all game animals from the water holes during the long dry season. To reach water undefended by tribesmen the antelopes and zebras must go outside the park, into territories where they have no legal protection and where they are slaughtered eagerly by natives and sportsmen alike.

Of all the African tribes, the Masai are least likely to attack game animals directly. Often these people take a fatalistic attitude when a lion makes off with a calf. But their cattle leave less grass and water available to any other creature. Their sheep and goats increase the devastation. Under millions of hoofs the grasslands of the Serengeti National Park become dust and soon will match the deserts farther north—deserts unlike those in North America in that they shelter no living thing.

Today Martin Johnson and his explorer-wife Osa would scarcely recognize the African plains over which they flew in early airplanes, photographing the living carpet of wild-life undulating below. From the more convenient light air-craft of modern times they would see only a land of grow-ing aridity, etched in every direction by stock trails, studded with thorn scrub only a rhino or a giraffe would appreciate. By default the desert succulents shunned by cattle increase and spread. Herbage useful to native antelopes cannot com-pete while grazed and browsed and trampled by so many plant-eaters year after year.

Currently the prestige herds of the Masai greatly exceed the carrying capacity of the park, even if all game animals were obliterated. A reduction of about forty per cent is needed immediately just to maintain the cattle, sheep, and goats. To such a drastic move, as also to evacuation of the park by all human beings and their livestock, the Masai are understandably resistant.

To many Africans the mere presence of wildlife is a re-minder of how long the Dark Continent remained a primi-tive place, its people far behind the rest of the world. Few would raise a hand against eliminating all wildlife, both in-side the parks and out. A long period of education would be needed to bring Africans to the point of view of the late King George VI of Britain: "The wild life of today is not ours to dispose of as we please. We must account for it to those who come after." Africa's wildlife may be gone com-pletely before education can reach a significant number of the native people.

Anywhere in the world, pride in the presence of domestic

livestock in excessive numbers "goeth before destruction" of the native wildlife, the plant cover, the soil, and the herdsmen too. Some of the great deserts seem to have come into existence this way.

I N *1740 the naturalist J. G. Gmelin recorded a legend . . .
told him by the Tartars. These Oriental tribesmen had
passed the story from one generation to the next for centuries. It concerned a small bird with short, rounded wings
—the Corncrake of Europe and Asia. At the proper seasons,
the legend claimed, each Corncrake sought out a friendly
Crane. The larger, stronger bird took the Corncrake on its
back and flew off with its passenger. Only thus, the Tartars
told Gmelin, could so feeble a flier as a Corncrake travel
from wintering areas to breeding grounds and return each
year. The Tartars underestimated the Corncrakes' vigor and
persistence, and overrated the amiability of Cranes. But
many other creatures do ride along without much effort on
their own part.*

LORUS AND MARGERY MILNE: *Paths
Across the Earth* (1958)

†◊

CHAPTER · 12

Sequels to Spread

❀

T HE United States acquired a new bird, the cattle egret,
in 1952. Small flocks of this little white heron arrived from
South America and made themselves at home. Just two decades earlier their ancestors had somehow flown unhelped
across the Atlantic Ocean, reaching the Guianas from an unknown embarkation point in the Old World—somewhere
between Portugal and Ghana, Madagascar and the Cauca-
sus.

147

Throughout the Old World the cattle egret is known as the buff-backed heron for its color at mating time, or as the tickbird from its habit of associating with large grazing animals and occasionally snatching engorged parasites from their sides and backs. In America the immigrant found a corresponding place among the feet of domestic cattle, catching insects frightened from the grass as the herd moved along.

Evidently cattle egrets are tolerant and observant birds, ready to benefit from any activity that makes insects and worms easily available. In Puerto Rico, an island the cattle egrets reached at about the same time as they reached New Jersey and Florida, we watched them following in the fresh furrows behind a tractor-pulled plow. As unafraid as though they had always lived there, the slender, graceful birds scampered after the machine, their snowy plumage contrasting sharply with the damp, dark earth.

No adequate census of insects is at hand to indicate what effect the cattle egrets may have on their favorite foods. Yet, to date, no one has seriously contended that the immigrants are besting any of our native birds in a competition for bugs and places to nest. How far the recent arrivals will spread and what their eventual impact will be are questions a single decade has not sufficed to answer.

Shortly after cattle egrets reached America, a creature that has been part of the New World scene since the Ice Ages or earlier made a sudden advance. Two decades have not allowed a new balance to be struck. Nor is man willing to let nature take her course, for his own welfare is at stake. The creature involved is "neither fish nor flesh nor good red herring," but the jawless lamprey, a dim-eyed denizen of the ocean now firmly established in our largest lakes.

Each lamprey leads a double life. First it is a flattened creature about the shape of a willow leaf and feeds on microscopic plants and animals in the murky waters of a swamp. Then it transforms into a new guise resembling a length of flexible rubber hose cut obliquely at the head end and flattened slightly into streamlined fins at the other. No paired fins of any kind develop. A row of gill openings, like portholes, along the sides look after respiration.

At the time of transformation, each little lamprey weighs about an eighth of an ounce and is nearly six inches long. Within thirteen to eighteen months, if living conditions are right, it will reach a third of a pound and some fifteen to sixteen inches from snout to tip of tail. During those months it will have attacked and killed at least twenty pounds of fish. Often these are fish in which man is financially interested.

Over most of the world inhabited by lampreys, they follow their transformation by an amazing downstream migration to the sea. There the creature's later life is spent, taking advantage of the big fish the ocean holds. Only as death approaches does a lamprey cease feeding, shrink in size, and seek out the stream of its youth. In riffles of fresh water near a suitable swamp, each attempts to find a mate and reproduce.

For years two exceptions were famous: the seemingly landlocked lampreys of Lake Champlain between New York state and Vermont, and an even larger race whose members did not go down the great rapids of the St. Lawrence River but instead spent their transformed lives finding fish victims in Lake Ontario. No one could be sure when the ancestors of these lampreys reached the lakes or what led them to give up long two-way migrations and a saltwater experience. The presence of lampreys in lakes so far from the sea may be proof of a different pattern of drainage toward the end of the Ice Ages. Lakes Champlain and Ontario may have been arms of the sea which became fresh water as the glaciers melted.

Scarcely anyone suggested another possibility: that lampreys beginning to ascend the St. Lawrence had affixed their tooth-studded sucker mouths to the bottoms of upbound ships in the hydraulic locks, bypassing the fierce rapids. A lamprey could have held on long enough to make the trip as a non-paying passenger all the way from tidewater at Montreal to lake water six hundred feet higher at Alexandria Bay or Kingston. Canals permitting such a passage were ready for traffic in 1825 as the first precursor of the present St. Lawrence Seaway.

Perhaps a lamprey, newly transformed from its filter-feeding youth to its more mature blood-sucking form, does not head for the ocean until it is hungry and, for lack of suitable prey, unable to try out its new sucking muscles by clamping its round mouth to a fish while rasping through the scales to a blood meal. The large fishes swimming in big lakes might suffice to obliterate the migratory habit of lampreys quite simply.

Size of victim is clearly important to a lamprey. Seldom do these external parasites attack a fish less than fifteen inches long. Other vertebrate animals of suitable dimensions are rarely met in lake waters where lampreys live. These "round-mouthed eels" have attempted, however, to fasten themselves to the grease-covered bodies of long-distance swimmers in Lake Ontario, and in 1952 were credited with hampering Miss Shirley Campbell of Fergus when she tried to break the women's record. Scars from lamprey attacks on whales far out at sea show that a lamprey recognizes no upper limit in size of victim so long as blood is available below its skin.

We may never know whether lampreys were in Lake Ontario prior to 1825. But it seems clear that neither the parasites nor their fish companions in this lowest of the Great Lakes could surmount the 160-foot plunge of Niagara Falls and reach Lake Erie, gateway to the remainder of the fresh-water chain. Another land-locked denizen of salt water, common in Lake Ontario, actually led the way. Within a few years after 1829, when the Welland Ship Canal was opened to take boats around the spectacular obstacle, lake herrings appeared for the first time in Lake Erie. Fishermen there and later on Lakes Huron and Michigan welcomed this addition to potential income. To catch lake herring these men used the same type of equipment that salt-water fishermen employ when the identical fish is known as an alewife or sawbelly.

The lampreys showed no important change in habits or spread until a century after the opening of the Welland Canal. After 1920 an occasional individual appeared above Niagara Falls in Lake Erie, but no colonization followed. Apparently Lake Erie's shallow, warm waters were a hurdle. The feeding streams entering this lake made poor sites for spawning and development of young lampreys. Fishermen of the upper lakes—Huron, Michigan, and Superior—scarcely knew the word "lamprey." They enjoyed the returns on a fresh-water fishery worth three million dollars annually at pre-war prices.

Lampreys entered this idyllic situation as villains in the early 1930's by finally making the full trip. They started colonies in the cooler water and fine tributary streams of Lake Huron. Within four years they spread into Lake Michigan, and each season saw them progress nearer its Chicago end. The commercial catch of lake trout, most valuable of the fishery assets, fell alarmingly. On the United States side of Lake Huron the catch had averaged 860 tons annually for forty years. Between 1936 and 1948 it dropped to 2.5 tons. In Lake Michigan the catch decreased between 1944 and 1949 from over 3,000 tons annually to 162 tons and continued downward.

Lake sturgeon virtually vanished, and with them an important sport fishery. The lampreys turned to suckers, walleye pike, and whitefish. They attacked lake herring

and, when these grew scarce, transferred attention to the "trash fish" of almost unsalable kinds. Only the small cisco and other little fish now prospered. The trout and pike and whitefish that formerly had fed on them were no more. But who would buy such puny fish? The fishing industry fell to pieces. Unused boats, nets, and maintenance sheds rotted beside deserted piers.

As lampreys in Lakes Huron and Michigan reached their peak numbers, many beaches grew unusable because of dead fish cast up by wave action. On most fish one or two great sores showed where lampreys had applied their mouths, rasped through the scales, and taken a fatal meal of blood and liquefied flesh. Fungus and disease had finished the attack on other fish that had escaped the lamprey, yet been unable to heal the wound the parasite had made.

No one was particularly surprised when herring gulls increased in numbers. Temporarily, at least, these scavengers could feast. But when the lampreys had done their work and few fish remained for gulls to find, bird-watchers expected a crash in gull populations. Instead, these birds remained abundant and even grew more numerous. How could fewer meals of fish improve their lot?

The solution to the gull puzzle lay in a research report from the University of Michigan Biological Station and the University of Illinois Zoology Laboratories. According to Dr. Lyell J. Thomas, nestling herring gulls are fed on regurgitated fish, brought partly digested by the parent birds. Ordinarily, whitefish and lake herring are the fish most easily obtained by adult gulls at nesting season, and provide a major part of the protein fare for the young birds. The flesh of these fishes is nourishing enough, but it usually contains the resting stage of a tapeworm, *Diphyllobothrium oblongatum*, to which young gulls are particularly susceptible.

Curiously enough, the parent gulls are free of these tapeworms and so is any fledgling as soon as its feathers have grown enough to keep the bird's blood heat high and constant. Only the fuzz-covered hatchling gulls are endangered, but often they succumb to an excess of internal parasites. Apparently, when the lampreys cut into the breeding stock of whitefish and herring, they forced the parent gulls to find other food for nestlings. And with fewer internal parasites, more young gulls reached flying age, swelling the population tallied by the bird-watchers.

The first lamprey in Lake Superior was recorded in 1946, but until 1953 this great body of fresh water still offered fishermen a livelihood. Lampreys in numbers had not yet ridden ship bottoms through the canals and locks by-passing rapids on the St. Mary's River. To prevent this translocation, United States and Canadian fisheries men tried all manner of equipment in the locks, hoping to clear lampreys from each ship before it passed to the next level higher. In spite of everything, the pests got through and began their spread in this final sanctuary. At its western end they would be twenty-two hundred miles from the salt water of the Atlantic Ocean from which their ancestors had come.

As hopes of preventing lampreys from using the locks faded, the scientists changed their tactics. Could the parasite be made an asset instead of a liability? Would the public buy its white, tasty flesh? But lampreys proved to be indigestible, and someone recalled that Henry I of England had died in 1135 of "a surfeit of lampreys." Even the carnivorous fish, such as lake trout, would have nothing to do with lampreys, large or small.

As people would not buy lampreys, the fisheries biologists began hunting for vulnerable points in the lamprey's life history. Perhaps the fasting adults, with only a few days left to live, could be dissuaded from entering streams where they could mate and reproduce. Engineers devised electric fences as barriers across the mouths of streams, and in 1956 the United States and Canada invested nearly three million dollars in supposedly lamprey-proof fences on all spawning streams flowing into the upper Great Lakes.

It is still far too soon for any decrease to be noticed in adult lampreys as a result of electric barriers. Progress may be expected in 1964 and following years, but not earlier. In the meanwhile, larval lampreys from fertilized eggs deposited in pre-barrier years will be maturing in upstream swamps and drifting into the lakes to take up their blood-sucking stage there at the expense of whatever large fishes they can find.

The time lag for control by means of electric barriers was so disheartening that biochemists set to work to test the gamut of poisons on swamp-caught young lampreys, searching for something with toxicity unique to these larvae. Between 1951 and 1959 better than six thousand different substances were tried, always with the hope that a low concentration of one of them would prove acutely poisonous to larval lampreys and harmless to other creatures inhabiting the same stream environments.

One of the chemical compounds proved effective. It was 3,4,6-trichloro-2-nitrophenol, soon trade-marked "Dowlap" by the manufacturers. An aqueous solution of the sodium salt of this substance killed all lamprey larvae within sixteen hours when tested at a concentration as low as twelve parts per million. Yet exposure to this poison for a whole day at three times the concentration caused no evident harm to rainbow trout, brown trout, brook trout, rock bass, bluegill sunfish, lake chubs, northern creek chubs, crayfish, or an assortment of insect young. Bullhead catfish proved susceptible to twenty parts of Dowlap per million, and mortality was found in yellow perch and white suckers when the concentration reached levels above thirty-two parts per million.

Results with Dowlap and other related nitrophenols are so encouraging that the Great Lakes Fishery Commission decided to stock Lake Superior with nearly a million young lake trout in the spring of 1960 and to follow this with an extension of chemical treatment and restocking in Lake Huron and Lake Michigan. Full control of lampreys need not be achieved before restocking becomes worth while, for the young of commercially valuable fish have several years to grow before they become subject to lamprey attack. Planning ahead at this time is regarded as a good "businessman's risk," one that may pay off handsomely if the lampreys can be kept curbed.

To many people it is still amazing that, by following natural rivers and waterways built by man, the sea lampreys have progressed westward halfway across the continent and occupied essentially all lakes with commercially valuable fish. Less than three decades were needed to turn a local curiosity into a curse. Now the deed is done, and the principal question left is whether the fisheries scientists can hold down the number of lampreys, letting fish and our only important inland fishery recover as a source of human food.

On the Pacific coast, America has reason to be apprehensive of the spread of another animal. This one, on land, has already devastated gardens in our outlying state of Hawaii. Its menace remains as an unwanted heritage from Japanese occupation of the Pacific islands, although the story is as incongruous as Aesop's tale of the turtle that won the race. On many bits of land in the South Pacific, the scars of World War II have healed. Yet the edible snails left behind have proved more destructive than all the bombs and artillery shells.

Admittedly the snail *Achatina fulica* is a big one, with an attractive brown shell marked with streaks of snow white, pink, green, or purple. It can fill the hand as an orange might, and may weigh up to a pound. Part of this weight, of course, is the sturdy shell into which the animal can withdraw its soft, slime-covered body if danger threatens.

When very young or old, the snail feeds on decaying vegetation. At intermediate ages it grows rapidly on a nocturnal diet of living plants, supplemented by everything from human excrement to dead members of its own kind. Within six weeks each individual matures as a male, only to change shortly to the opposite sex and commence laying eggs in batches of as many as three hundred, month after month. Each egg is the size of a large pea. Usually they are secreted under boards or in loose topsoil. Hatching may come within a few hours after the egg is laid, for the parent commonly delays deposition, giving the egg extra protection in this way. At a conservative estimate, a solitary gravid individual could give rise to nearly 1,100 million offspring in five years. At a pound apiece, this is more than five hundred thousand tons of mollusks, representing a frightful toll on vegetation.

Achatina fulica first became known to science as a denizen of islands in the Indian Ocean. Mauritius may have been its original home, although snails of this kind have fitted into the fauna of eastern Africa from Natal to Somaliland. On the continent's west coast they are common at least as far north as Ghana. Over much of this territory and on Madagascar their most persistent enemy is man, for the native people prize the snails as food and find all manner of uses for the shells. A civet cat, a land crab, one or two carnivorous snails, and a few kinds of beetles prey on young *Achatina* before the shell has grown too heavy and the creature is able to flood its surface too copiously with mucus. Together, they keep *Achatina* in check.

The earliest recorded big move for the East African snails came in 1847 when Mr. W. H. Benson, a British conchologist, visited Mauritius. Mr. Benson was struck with admiration for the huge snails he saw creeping about the garden of his host and fellow shell-collector Sir David Barclay. What could be more normal for the traveler than to pick up a few live specimens and tuck them into his baggage as he left, as souvenirs of Mauritius? But when he reached Calcutta, Mr. Benson realized that he could not carry the snails with him all over India. With tender care he released them in the Chouringhie Gardens just outside the city.

That part of the world has a density of human population greater than five hundred to the square mile, and starvation is more common than good nutrition. Meat in particular is lacking in almost everyone's diet there. Yet the big edible snails remained unacceptable as food. Instead, they competed with the sacred cattle for vegetation that might well have bolstered human meals.

Whether *Achatina fulica* received more help in spreading during the next half-century may always remain unknown. Certainly they reached Ceylon about 1900 and quickly became a serious agricultural pest. By 1911 they had become familiar all the way to the tip of Malaya, around Singapore. Seventeen years later they island-hopped, with human aid, to Borneo. There in 1931 a bounty was offered for snails and their eggs. Despite the destruction of five hundred thousand snails and twenty million eggs within a two-week period, the remainder seemed as numerous as ever.

In 1933 these mollusks reached Sumatra and Java, where they attacked young rubber trees. In the same year descendants of the Malayan stock spread to Japanese-controlled Formosa by way of Amoy, directly opposite on the Chinese mainland.

On Formosa the Japanese people welcomed the new arrivals both as an interesting food and as medicine of special virtues. Enthusiastically they sent living specimens to the home islands, innocently or deliberately circumventing laws forbidding the introduction and culture of a mollusk already famous for injuring crops in Ceylon and Malaya. Fortunately, the winters of Japan proved too severe for *Achatina*, even though it often avoids inclement weather by burrowing into the soil, using the earth as an overcoat.

No natural barrier of cold protected the Hawaiian islands when descendants of *Achatina* on Formosa were brought in 1936 by a Japanese visitor from Honolulu. The snails are not only out of control on Oahu, but have spread also to Maui. After many trials the Hawaiian Sugar Planters' Association reluctantly concluded that "eradication is impossible without the expenditure of vastly more money than the Territory can afford."

The real explosion of *Achatina* populations came as the Japanese took their new food with them through the Pacific islands. A start was made as early as 1938 on islands mandated to Japan after World War I. About May of that year they were introduced from Japan on Palau in the Caroline group, and quickly propagated to great numbers. Two Japanese biologists, Drs. Teiso Esaki and Keiso Takahashi, urged the government agents on Palau to forbid the rearing of the snail there, and attempts were made to exterminate it. They were fruitless. The situation soon was duplicated in the Bonin Islands, on Ponape, Yap, Tinian, Rota, and Saipan in the Micronesian group.

A dozen years ago, when the U.S. Navy was sent in to straighten up the islands of the Pacific Trust Territory, billions of snails had taken over much of the Dutch East Indies and many islands of the Philippines, as well as New Britain, New Ireland, and New Guinea. They hid in salvageable war materiél that was being recovered and shipped to America. Thousands of specimens were scalded out of jeeps through use of live steam under pressure. Even larger numbers died on California docks when this treatment was given bulldozers and ambulances brought back from Pacific islands. Fortunately, no *Achatina fulica* has yet established a beachhead in America. But how long this can be prevented remains to be seen.

When *Achatina* proved to be out of control on Hawaii, Dr. Francis X. Williams of Harvard University was sent to East Africa in search of the native enemies of the snail. Perhaps one or more of them could be brought to areas of infestation and provide a natural counterbalance. But none of the enemies except man proved to be selective. For a time in the early 1950's scientists considered introducing *Gonotaxis kibzewiensis*, one of the carnivorous snails that attack *Achatina*. Living specimens were sent to Honolulu for study. Then the idea was discarded, for the *Gonotaxis* showed that it was equally ready to bore through the shell of any large snail on Hawaii—not just *Achatina*—and might easily exterminate some of the native snails without making significant inroads on the devastating immigrant.

Californians discussed the pros and cons of liberating *Gonotaxis* snails in their state before *Achatina fulica* can become more than a menace. Already gardeners in California have trouble with the European garden snail *Helix aspersa*, introduced many years ago. The *Gonotaxis* might help. Yet once more the preliminary tests indicated that possible gains would be too slight to outweigh the possible (though unknown) disadvantages.

The East African *Achatina fulica* is not the only one to fear. Its near relatives may be worse. In barely over a century, in the very garden on Mauritius from which *fulica* started out for Calcutta in a conchologist's luggage, the related *Achatina panthera* has been outfeeding and outbreeding *fulica*. The *panthera* may eventually take over the island altogether. Certainly Sir David Barclay, who owned the garden and introduced *panthera* as an interesting addition to the local fauna, foresaw no such gradual change.

A still larger member of the same group is the eight-inch Nigerian snail, *Achatina achatina*. Whether it can devastate vegetation in the same way away from home has yet to be seen, although living specimens have already been brought to the North American continent. A few years ago a reputable importer of wild animals, Mr. Warren E. Buck of Camden, New Jersey, bought several giant *Achatina achatina* to grace his collection. The United States has no nationwide legislation prohibiting entry of this or any other snail.

Shortly afterward a California shell-collector received by mail two living specimens of the same giant kind and generously offered them for display at the Steinhart Aquarium in San Francisco's Golden Gate Park. There Dr.

Tracy I. Storer, a member of the state's Committee on Alien Animal Importation, recognized them and started an investigation. California is one of the few states with local regulations covering snails. The *Achatina*, even though not *fulica*, were destroyed lest some accident free them in so productive a garden area.

Both the lamprey and the *Achatina* snail have spread as far as natural barriers have permitted. Both are merely following a fundamental drive in all living things, a constant pressing against boundaries. Man has shown this same characteristic consistently, in spite of civilized tendencies to settle down and merely improve the *status quo*. Barely had the idea of "one world" arisen, through man's recently gained ability to travel swiftly from any part of the earth to any other, when he recognized interstellar space as a new challenge to his spread.

Throughout his travels man has fine-combed the earth for plants and animals he could relocate and raise for his own benefit. For them too the concept of "one world" could be extended. But man's welfare and that of his domesticated following are seldom served by wide application of the same rule. We thrive best by keeping non-human life of most kinds as thoroughly isolated as possible, each in its own homeland where it is in balance with its neighbors.

Most creatures create problems in a new land. To date, the cattle egrets have proved exceptional. One can only be thankful that they demonstrated their surprising ability to travel so far without help at a time in human affairs when the move could be appreciated in its true perspective. And we are delighted that no sad sequels have so far followed their spread.

Often in the past a new human activity has unwittingly led to situations worse than the recent disasters occasioned by lampreys and *Achatina* snails. Perhaps the earliest at which history hints came at about the time when Alexander the Great is reputed to have sat weeping because there were no more worlds to conquer. His armies apparently brought back to Asia Minor both the house rat and the disease carried by its fleas: typhus fever. Malaria may have come along at the same time.

The Christians who trekked to Palestine on a series of Crusades to free Jerusalem from Moslem domination seem to have carried to Europe on their return not only a fresh curiosity about animals and vegetation but also the rat, and typhus, and perhaps malaria. Columbus and the Spanish expeditions that followed his route in search of gold to spend and souls to save certainly brought smallpox and measles as well as Christianity and rats to the New World.

The conquest of the Indian areas of modern Latin America by so few Spaniards would scarcely have been possible had malaria and yellow fever been in the Western Hemisphere at that time. Both diseases turned up in the years following introduction of Negro slaves from Africa, and the mosquitoes serving as carriers of infection may well have gone through repeated generations in the casks of fresh water brought on the slaving ships. How could anyone predict that a few small insects traveling across the Atlantic Ocean in company with dejected captive men and women could someday frustrate the French in their attempt to build a Panama canal?

Who could predict that the Welland Ship Canal would light the fuse to an explosion of lampreys, ruining fishermen a century later in the upper Great Lakes? What, we may wonder, can come throught the new St. Lawrence Seaway?

No one would have suspected that a snail could be the real winner in a modern war. Yet on Pacific islands as close as Hawaii, *Achatina* has regrouped mindlessly, ready for the conquest of America as well. Successful infiltration of our defenses seems possible at any time.

In Mauritius, Madagascar, and much of Africa, the *Acha-*

tina snails have no barriers to cross, and their presence constitutes a danger there as well. If the native people, who are making such rapid strides toward independence and adoption of white man's ways, change from use of snails as food and snail shells as utensils, the present adequate control on *Achatina* numbers may dwindle away. What then?

Man is probing the earth's last hidden corners for places where he can increase his food supply. Other creatures are also spreading into new localities and competing with him there. In this contest, human efforts succeed significantly through preventing our many competitors from spreading throughout the world. The longer their spread can be postponed, the more plant food man can turn to his own uses.

THE day after Flannery had counted the guinea-pigs there were eight more added to his drove, and by the time the Audit Department gave him authority to collect for eight hundred Flannery had given up all attempts to attend to the receipt or the delivery of goods. He was hastily building galleries around the express office, tier above tier. He had four thousand and sixty-four guinea-pigs to care for. More were arriving daily.

ELLIS PARKER BUTLER: *Pigs Is Pigs* (1906)

~❂~

CHAPTER · 13

Conquering Rabbits
❀

IN 1953 the landowners of England began to free themselves for the first time from an annual loss equivalent to a tax of more than two dollars on every man, woman and child in the country. The loss consisted of cereal crops never harvested, of grasslands producing less than normal yields of fodder, and expenditures to protect state-owned and privately held plantations.

These losses began inconspicuously some seven centuries earlier, soon after William the Conqueror compiled his Domesday Book, listing the country's resources down to the last ox and pig and stretched hide. Among those assets the Conqueror's barons missed one creature familiar to them in their native Normandy. To remedy the situation and to provide themselves with better hunting, they introduced the target they longed for—the European rabbit. It throve in the British Isles and soon seemed a natural part of the landscape it devastated.

Englishmen, whether of Norman descent or not, came to enjoy hunting rabbits and even developed the beagle race of dogs to help them in this sport. When Englishmen began settling Australia, they too missed seeing rabbits hopping gaily about. Thomas Austin, Esq., who owned Barwon Park near Geelong in the Australian province of Victoria, was a man of action who decided to improve the local scene and perhaps remedy the homesickness of some of his men. He ordered two dozen European rabbits from "home." They arrived aboard the Black Ball clipper *Lightning* on Christmas 1859 and, with Mr. Austin's blessing, went free four days later on Australian soil.

The blessing was superfluous. No doubt all of the doe rabbits were pregnant upon arrival. Rabbit gestation lasts a mere thirty-one days, and in southeastern Australia each doe can produce up to ten litters of at least six young annually.

Within six years Mr. Austin had killed twenty thousand rabbits on his estate. He estimated the breeding stock at an additional ten thousand. Probably his figure was low, as the rabbits had essentially no enemies. Later in this same area the rabbits demonstrated that even when two thirds of them were shot every year after the first, twenty-four rabbits could become 80,368 in three years, and twenty-two millions in six. If Mr. Austin harvested only twenty thousand rabbits, he barely nicked the population.

New Zealanders had been trying to introduce rabbits on both North Island and South Island ever since the 1830's. For some reason the animals did not survive. A new attempt in 1862, credited to a Dr. Menzies, the superintendent of Southland, succeeded and the rabbits spread in all directions. Within seven seasons they were "becoming a nuisance." Five years later landowners referred to them as "armies" attacking, dividing, capturing mountain ranges, debouching on arid plains, crossing rivers, and digging in as they progressed. "Hillsides were alive, pastures devastated, flocks starved, sheep farmers ruined, and hundreds of thousands of acres had to be abandoned." By then North Island was scarcely better off.

Neither Australia nor New Zealand had any native predators to hold back the rabbits. Some house cats went wild, but on the down-under continent itself they found the native marsupials even easier prey than rabbits. The Australians introduced foxes from Europe, only to discover that the foxes preferred the small, plump, rabbit-sized rat-kangaroos known as tungoos. Tungoos and rabbits had come to live together amicably, the former even occupying the mazes of tunnels the rabbits dug in each warren. The tungoos, however, reproduced slowly and, with foxes as competition, became a vanishing species.

The feral house cats increased in numbers and caught so many of Australia's unwary birds that insects were suddenly liberated from their flying enemies. Leaf-eaters and boring insects multiplied, attacking the native eucalyptus trees with unprecedented vigor. Among the bare branches the white settlers saw koala "teddy bears" exposed, each koala trying to find some foliage the insects had overlooked. To save the eucalyptus forests, the Australians began exterminating the koalas—shooting these slow-moving creatures wherever any could be found.

To rescue the koalas from oblivion, the Australian government established several sanctuaries, some on islands and some on the mainland. In these the koalas benefited from protection and began a slow increase in numbers. But the eucalyptus trees that are their sole food were too severely plagued by insects. They could not support koalas too, and soon began to die. In some areas the sanctuary wardens eliminated the koalas to save the trees. In others the koalas ate themselves into starvation and the refuges ceased to have any significance.

Gradually the European rabbits came to occupy more than a third of the continent in spite of hunting, poisons, traps, dynamited warrens, ferrets, foxes, dogs, and a supposedly rabbit-proof fence of wire netting stretched more than seven thousand miles across the country. The Australians exported 700 million rabbit skins and 157 million frozen carcasses in a single decade, supplying most of the world's demand for felt of rabbit fur and the ingredients for countless rabbit pies and rabbit stews. Rabbit-hunters had fun and made money. Nearly everyone else lost. Five rabbits ate as much as one sheep, and the combined payment for five rabbit skins and five carcasses amounted to less than a third of the value in one sheep's annual crop of wool, forgetting all gain from lamb and mutton, hides and tallow.

As recently as 1950 the Australians were counting rabbits as though in a nightmare. Just a little arithmetic with a pencil showed that the progeny of each pair, if unmolested, amounted in three years to about thirteen million individuals! Without making any real headway against the menace, the people were spending millions of pounds sterling on bounties, on poisoned apples, poisoned carrots, poisoned pears, poisoned quinces, and pollard (wheat bran) poisoned with arsenic. Specialists went around the country digging out the warrens, many of them mazes of tunnels to a depth of nine feet with interconnections covering as much as a quarter of an acre. Heavy equipment was wheeled into the fields and used to pump into the burrows lethal gases such as carbon monoxide, carbon disulfide, cyanogas, and chloropicrin. No other remedy could be found. The hunters not only had failed to control the rabbit plague, but also had become bored with so many targets of the same kind.

During the early months of 1950 bacteriologists and

wildlife managers for the Australian government worked together in testing a suggestion that had reached them from South America the previous year. Brazilian cottontail rabbits carry a disease virus described first in 1897 among animals near Montevideo, Uruguay. Although the South American rabbits are almost immune to the disease, the virus can be transferred from them to domestic rabbits, which are a special strain of European rabbits. For some reason the domestic strain showed tremendous mortality. Perhaps the South American virus would be effective with wild European rabbits in Australia.

The desperate Australians tested samples of the "myxoma" virus on both captive European rabbits and a wide assortment of marsupials. Only the rabbits were affected by it, and man was completely immune. By Australian winter the indoor tests ended. Each had been completely satisfactory. The next step was a field experiment. For this a number of European rabbits were caught in Australian fields, inoculated with the myxoma, and freed again in the rainy Murray-Darling drainage region of western New South Wales. This was close enough to both Adelaide and Melbourne so that scientists could keep careful watch on developments.

Spring brought an abundance of rain but no change in the rabbit population. The disappointed experts sent in their reports: experiment a failure. Summer began while the conclusions were being printed; it was December in the Southern Hemisphere. Suddenly rabbits began dying by the hundreds, by the thousands, all over the Murray-Darling drainage. At better than three miles per day, the disease spread up the river valleys in every direction.

Newspapers broke out with banner headlines in type sizes that had not been used since the end of World War II: EPIZOOTIC A GRAND SUCCESS! And everyone learned the name "myxomatosis" as well as the symptoms of the disease.

Rabbits sickened, became sleepy, while their skins broke out in great sores. Mosquitoes and other blood-sucking flies attacked, particularly around the sores, and got myxoma virus smeared over their mouth parts. Then these parasites flew to healthy rabbits and inoculated them in turn.

Myxomatosis gained momentum in January 1951, but in most areas the peak of rabbit deaths had passed by February. And by the end of March, when southern autumn arrived, the surviving rabbits were hopping about as though nothing had happened. Mosquitoes were the chief agents of transfer, making myxomatosis a summer disease. Gleefully the Australians looked forward to another try the following year. The only objections came from humanitarians who were offended by the sight of dying rabbits and from hunters, skinners, and meat-packers who relied upon rabbits for a livelihood.

When the mosquitoes began flying in the Australian summer of 1951–2 they encountered newly inoculated, freshly released European rabbits in practically every infested part of the country. Never before did a deadly disease sweep an animal population with such devastating effect. Out of every thousand rabbits in Australia, at least 995 succumbed to myxomatosis. Often the rare survivors ran about wildly, as though seeking company—and met a fox or a farmer with a gun. Grass sprang up where none had been seen in the memory of the human inhabitants. The sheep put on weight almost twice as rapidly, and pastures supported far larger flocks. Australian landowners began to dream of a continent without rabbits.

Such wonderful news spread throughout the world. In France a leading pediatrician, Dr. Paul Armand-Delille, saw myxomatosis as the long-sought answer to the European rabbits that had repeatedly attacked his vegetable garden. Quietly he sent to Australia for a supply of the virus. Before many months elapsed he followed the accompanying instructions, freeing several inoculated rabbits on his estate near Chartres in time for mosquitoes of the 1952 summer in the Northern Hemisphere.

Dr. Armand-Delille broke no law in importing the virus and using it on his land. Yet his action upset an age-old balance between the European rabbit and its native environment. Many other landowners regarded the animals as pests. But one Frenchman in every ten had a hunting license and used it to hunt rabbits, as almost no other game existed outside private preserves.

Rabbit meat was such a staple food among Frenchmen as to be priced by the government for inclusion in calculating the cost-of-living index, the basis upon which wages might be revised upward or downward. French furriers depended heavily upon rabbit pelts; though fully aware that the fur lasts only a twentieth as long as otter, they were sure that garments trimmed with rabbit would be out of style before they wore thin. For rabbit fur they thought up such euphemistic labels as "arctic seal," "clipped seal," and "polar seal." The word "lapin" used in America is merely the French word for rabbit.

European rabbits in France proved just as susceptible to myxomatosis as those in Australia. Even after the mosquito season ended, the disease continued to spread, apparently transferred from sick rabbits to healthy ones by a large population of fleas. Everywhere the rabbits died. The stench from their decomposition was unbearable. Licensed hunters were aghast, the manufacturers of guns and ammunition despondent. Quickly a new society was formed, the Association de Défense contre la Myxomatose, with members all over the country. They tried vaccinating rabbits against myxomatosis. Yet the disease spread faster than ever. Furriers went out of business. The Assembly met in special session to consider the economics of the emergency: What would France be without rabbits? Would everyone strike for higher wages? Should rabbit meat be omitted in calculating the cost-of-living index? What other meat could a poor Frenchman afford for his family?

The despair in France was balanced only by the happiness in Australia. But France is not a continent, and myxomatosis respects no arbitrary boundaries. The disease spread northwest into the Baltic countries and Germany, and across the mountains into Spain and Italy. There the disease was more welcome, for rabbits had less place in the food budget. They were kept in cages, far from any wild animals, as a completely domesticated food supply. Italy still has between six and seven million adult domestic rabbits maintained in this way.

Game wardens on the big estates in the south of Spain began to worry. On these preserves live the last of Europe's true lynxes, and the big cats dine regularly on European rabbits, with young deer a second choice. The owners enjoyed having the lynxes survive, but wanted deer to hunt. That lynxes would kill dogs, foxes, and mongooses without eating them had seldom troubled the owners. But if the lynxes could find few rabbits and turned to deer, would it be necessary to destroy Europe's remaining big cats? The problem seemed even more acute because Australian eucalyptus trees had been introduced recently. Almost shadeless forests of eucalyptus were crowding out the native trees and providing little cover for lynx or deer.

The English Channel is but twenty miles wide, yet the conservative English did not help myxomatosis cross it until 1953, when France seemed about to run out of rabbits— both healthy and diseased. Finally the deed was done anonymously, perhaps by more than one person. Animals dying with myxoma virus appeared in the county of Kent during October. Through the mild winter the disease turned up in eleven other places in Essex, Kent, Suffolk, and Sussex. And with the emergence of blood-sucking insects in the late spring and summer of 1954 it spread rapidly. By December the rabbits were declared doomed in all but seven counties of England, fourteen counties of Scotland. The disease had been carried to many of the outlying islands as well.

Officially it is still illegal to deliberately aid the spread of myxomatosis on the island of Great Britain. But this concession to the humanitarians was never enforced. Sportsmen tried in vain to get government support for protecting rabbits as targets, but landowners shouted them down. Dogbreeders in Wales complained that they could no longer train dogs for export, without rabbits for the dogs to chase. Game wardens and wildlife specialists merely kept count. They saw one hundred million rabbits in the British Isles shrink to less than three million by 1956. Crop yields leaped upward as much as fifty per cent. The grazing season lengthened. Farmers claimed a saving of at least forty-two million dollars a year. And when a strain of rabbits in Nottinghamshire appeared to be resistant to myxomatosis, swarms of rabbit-exterminators were rushed in to prevent any resurgence of the pest the Normans had introduced so long ago.

Rabbits surviving myxomatosis were appearing in Australia too. Wildlife officers could not learn immediately whether some strains of rabbits had a natural immunity, or whether the myxoma virus itself had mutated into less virulent forms. But with complete elimination of rabbits so close, no chances were taken. More animals, inoculated with the most virulent type of virus known, were released to reinstate the disease wherever wild rabbits remained. At the same time the ranchers, farmers, and government officials hurried forth with tons of poisoned fruit as well as mechanical equipment with which to rip open rabbit warrens and get any survivors out to where they could be shot.

The foxes introduced into Australia to control rabbits came in for criticism. Perhaps they too should be eliminated. Foxes seemed to prefer rabbits that were coming down with myxomatosis and were too sick to run far. This reduced the spread of the disease, as the fox often killed the rabbit before its open sores developed and before mosquitoes or other blood-sucking insects had had a chance to become contaminated. Unless the mosquitoes did their share, myxomatosis failed to spread.

To dodge this difficulty, conservation officers began putting inoculated rabbits in cages of chicken wire through which blood-sucking insects could go. This method kept the rabbits safe from foxes until local mosquitoes and black flies and sand flies could become contaminated. A careful publicity campaign was necessary to alert Australians to the reason for preferring rabbits to sicken and die in caged confinement rather than fall victim to hungry foxes.

Since the spread of myxomatosis depended upon having infected, sore-carrying rabbits where mosquitoes could find them, it became important to know more of the feeding habits of Australia's insects. Entomologists began trapping mosquitoes alive and obtaining serological examinations of any that had had a blood meal. In this way it was possible to learn which mosquito had bitten what warm-blooded victim. Again the public was astonished. Imagine anyone wanting to know what a mosquito had for dinner!

A common mosquito of the river valleys, *Anopheles annulipes*, seemed attracted equally to rabbits and to cattle. If given a chance, it would attack man, poultry, and other animals. It could carry myxoma virus from rabbit to rabbit, or malaria from man to man. Yet this *Anopheles* seemed to bite man too rarely to be important in transmitting malaria. Hence it should be regarded as beneficial in furthering the spread of myxomatosis. The idea of a beneficial mosquito was rather new. Many people found it hard to accept.

Another mosquito, *Culex annulirostris*, ranged widely, biting all sorts of warm-blooded animals, whereas the Australian race of cosmopolitan *Culex pipiens* seemed to alternate between rabbits and poultry. When either of these mosquitoes bit rabbits, they could transmit myxomatosis. They were "good mosquitoes" too. Yet both of them were suspected of bringing to man the dangerous type of sleeping sickness known as Murray Valley encephalitis. Entomologists wondered whether it would be wise to encourage increased numbers of the three beneficial kinds of mosquitoes, as improvements in the transfer of myxoma virus might be paralleled by the spread of human diseases. Only about *Culex fatigans* were they certain. This mosquito came indoors and bit man savagely, and there was no indication that it played a helpful role in controlling rabbits. It was a "bad mosquito." Never before had so much been known about the food habits of these insects.

In England and Scotland, fleas rather than mosquitoes served in the transfer of myxomatosis. Fleas have no wings and must depend upon their host animal coming within leaping distance in order to take up adult life among a rabbit's hairs. To find a mate, a flea may have to shift from one rabbit to another—when the rabbits are close together. And even when successful, a mated flea does not lay eggs on the rabbit, for fleas in the immature stages of their development are not parasitic but scavenge among the grass blades and root tangles. Each new generation of fleas must find its own rabbits.

If rabbits are numerous, fleas have no difficulty completing their life cycle. But where rabbits are scarce, many fleas starve to death before they can find a host. Rabbit fleas threatened to become extinct where myxomatosis was successful. Yet without their help the myxoma virus might not pass from a sick rabbit to a healthy one. Myxomatosis might not reach potential victims.

On some of the rabbit-infested offshore islands, myxomatosis did not spread at all from inoculated rabbits released to start an epidemic. A lack of fleas was blamed. British entomologists began raising rabbit fleas in captivity, so that each inoculated rabbit could be furnished with a reasonable number and ensure propagation of the disease.

Both in Britain and on the continent of Europe, rabbits became so scarce that foxes changed their habits. Poultry-raisers had to be far more careful to protect their flocks. Yet the foxes showed no signs of malnutrition. They simply altered their diets, eating more plant foods and improving their skill at catching rats and mice. Country people and city naturalists alike reported a phenomenal decrease in rats. The populations of field mice and voles shrank too, and owls felt the change. In one thousand-acre area near Oxford, studied by the eminent British naturalist Dr. H. N. Southern, the number of young tawny owls raised in the deciduous woodland dropped from an average of twenty to just under three in 1955. Stoats and weasels, which depended upon rabbits more than most other predators, decreased sharply in many areas.

By 1956 a new balance was struck. The European hare, which is immune to myxomatosis, became comparatively common and ate a small part of the food formerly used by European rabbits. With fewer stoats and weasels, the mice and voles increased again. The number of tawny owls breeding in Dr. Southern's woodland reached levels higher than any he had recorded before. Foxes got more meat and ate less plant food. Scientists watched, fascinated by the "lateral cross-connections between the consumer layers" in the animal community as Britain's wildlife adjusted itself to the virtual absence of rabbits.

Crop yields continued to increase. An orchid long believed extinct in Britain reappeared, spreading out of a place of concealment. Wildflowers previously considered rare became almost commonplace: they must have been favorite foods of rabbits. The public found plenty of entertainment, even excitement, in watching farmers rush around with government agents, working day and night to release more inoculated, flea-carrying rabbits, to spread poison bait or fumigate with cyanide gas, to set traps of every kind, or snares, or long nets. The landowners used ferrets, dogs, rifles, shotguns, and power equipment for destroying any inhabited burrow or cover in which a surviving rabbit might hide. Still a few remained.

In France the forests began to regenerate—a phenomenon no one alive had ever seen before. Holland's sand dunes ceased to be a serious problem because dune-fixing vegetation now spread dramatically over the surface, protecting it from wind. In Australia, meanwhile, grass grew in regions classified as deserts. Fields that had barely supported one sheep now fed two cows quite generously. The gains were clear. Yet a good many rabbits still hopped about the down-under continent. Australians considered introducing European rabbit fleas to help the native blood-sucking insects. Unfortunately, the fleas proved willing to live also on some of the local marsupials, and the idea had to be discarded.

To really rid any area of rabbits, more than myxomatosis seemed necessary. Yet the idea of being free of rabbits held great appeal. People on New Zealand and on the island of Tasmania, off Australia's southernmost bulge, listened attentively and enviously to the glowing reports from Australian farmers and ranchers, as well as from their counterparts in Europe and Great Britain. For some reason myxomatosis had failed to take hold in New Zealand and Tasmania. The landowners and the government there decided to plunge into an intensive extermination campaign anyway. But how should they begin?

One possibility lay in finding a more potent poison. An American product known as "1080" held great appeal, even though its use in the United States is illegal except in the hands of licensed pest-control officers. This white powder is sodium fluoroacetate, a substance for which no antidote is known. A solution of only half an ounce to a gallon of water is strong enough to poison an apple or a carrot dipped in it. Captive rabbits seemed unable to detect the lethal coating.

To use "1080," Australians, New Zealanders, and Tasmanians moved their livestock out of area after area, then spread poisoned food in long plowed furrows and waited for the rabbits to enjoy their last meal. Afterward the furrow could be closed again, burying the hazard so deep that domestic animals would not find it, and letting the leaching action of downward-sinking rainwater eventually dissipate the danger altogether.

In New Zealand, where introduced deer are also a pest for which bounty payments are offered, the government sponsored the broadcasting of poisoned food from airplanes in mountainous and unpopulated areas. At the same time large-scale activity on the ground, using arsenic and phosphorus poisons, supplemented the more customary use of traps, nets, fumigants, guns, and warren-ripping equipment. Legislation wiped out the traffic in rabbit skins and meat by making it illegal. Perhaps more important was a progressive transfer of the responsibility for controlling rabbits from the individual landowner to locally elected "Rabbit Boards," toward which both the landholder and the government contributed tax funds.

One surprise came by 1957. Not only did fewer rabbits mean more luxuriant grass in New Zealand. Taller grass also meant fewer rabbits! The European rabbit "colonizes bareness" and thrives on depleted, eroded soil. The animal finds its favorite annual plants growing on almost bare ground or in close-cropped turf; there it becomes a nuisance. It is favored by close grazing, and hence by the presence of too many sheep and cattle on the land. Drought is the rabbit's friend, and a good grass fire helps it considerably. It seems that merely following known methods for improving the land may eliminate the rabbit problem.

New Zealanders began to realize that the prime offender was not rabbits after all. It was the combination of rabbits and excessive numbers of domestic livestock. If the land was fallowed—used for grazing only on a rotational basis— the rabbit warrens would disappear. Every measure improving the soil, even reseeding it with good grasses, was a blow to the surviving rabbits. And by 1958 the New Zealanders were crowing over their accomplishments, pointing to their lands as examples the Australians might envy—even if myxomatosis had been no help in ending a century of European rabbits.

History, aided by rabbits, has repeated itself at many places. On various islands of the South Pacific, green paradises have become deserts after less than a century of rabbits. This native of Europe seems to be capable of living almost anywhere: to within two degrees of the equator in central Africa, and on the frigid, windy land of Tierra del Fuego in latitudes comparable to those of Labrador.

The rabbits of Tierra del Fuego started from a small number of domestic animals released about 1910 on the Chilean half of the main island. By 1930 they had become numerous and spread inland. By 1947 they reached plague proportions despite all control measures, and the sheep industry there showed heavy losses from malnutrition. A realistic approach to stocking the sheep ranges might improve the situation, but the islanders question whether they could themselves survive the economic pinch if still fewer sheep were pastured on the bleak hills.

An interval of twenty years between the time of successful introduction and the first estimate of rabbits as "common" is a pattern seen also in the Hebrides, a group of five small, partially connected islands just northwest of the Scottish highlands. There the thousand acres of pasture land had supported about one head of livestock per acre until the 1930's. Then the combination of heavy grazing and rabbits led to soil deterioration. By 1951 only about fifty sheep and twenty-five young cattle could find enough to eat on the whole area. Myxomatosis was introduced in both 1952 and 1953, but, despite the presence of rabbit fleas, the disease did not take hold. In 1954 neither myxomatosis nor any decrease in rabbits was evident, and the Hebrides were largely written off as a pasture area for livestock.

Ever since the turn of the century America has felt fairly safe from foreign birds and mammals that could become pests in the Western Hemisphere. The sentiment against English sparrows and European starlings was so strong in 1900 that Congress had no difficulty in passing the Lacey Act forbidding any further introduction of warm-blooded animals. But the law has one loophole: it says nothing about transporting European rabbits from one United States point to another. And European rabbits were already in the country by 1900, both domesticated and wild.

The wild ones are descendants of domestic rabbits released about 1900 by the lighthouseman on San Juan Island in Puget Sound, near Seattle, Washington. He hoped to enjoy rabbit meat as food that would cost him only the trouble of catching the animals. Through his help the rabbits colonized San Juan Island and also little Smith Island, an area of some fifty square miles.

By 1924 the lighthouseman's successor had begun begging the United States government for assistance in controlling rabbits. A rough tally on Smith Island showed more than thirty rabbits to the acre, and almost no vegetation except bracken fern, cheat grass, and tarweed. Rabbit burrows so honeycombed the island that foundations of buildings were undermined and collapse seemed imminent. Coastal bluffs already had begun to cave in, gradually shrinking the size of the island itself.

179

San Juan Island has a good many dogs, cats, raccoons, and mink. Yet these do not control the rabbits. Interlacing tunnels provide a secure haven, and farmers' attempts to poison rabbits without endangering pets and domestic livestock meet with indifferent success. In 1954 only one man seemed to be making much use of the abundant, nocturnal rabbits. With a jeep and a long-handled scoop net, he caught them at night, and claimed to have shipped out more than fifty thousand annually for many years. He grumbled at the poisoning campaigns, but had no comments about the damage the rabbits did.

No one has yet released on Smith Island or San Juan Island any domestic rabbits inoculated with myxomatosis. Strangely, there has been no campaign to wipe the European rabbit from the Western Hemisphere. Even more incredible is the fact that since 1950 the members of beagle clubs in state after state have clamored for release of rabbits from San Juan Island. They are anxious to dodge the wise provisions of the Lacey Act and get European rabbits into the farms and fields of the continental United States.

In Ohio the state wildlife officers tested San Juan rabbits in pens and found them just as ready to burrow extensively as the European rabbits in England, Australia, or New Zealand. The test was repeated in Maryland in a half-acre pen enclosing a good variety of woody and herbaceous plants. For over a year the population was held at just three rabbits. They ate clover and honeysuckle, locust sprouts and roots, and girdled the bark on every apple and wild-cherry tree. No gain could be seen from introducing into Maryland as a sport animal a creature with nocturnal habits, especially a rabbit digging in so deeply and liking so many valuable plants. Yet by 1955 European rabbits from San Juan Island were indeed liberated on the mainland: an unknown number in Wisconsin, almost five hundred in Indiana, "several hundred" in Michigan, several thousand in Ohio, and "some" in Pennsylvania.

Whether any of these deliberate releases, each planned to circumvent the Lacy Act, has led to colonization is not yet known. New Zealanders tried repeatedly for thirty years before the introduction was deemed a success. Nor is there any agreement on the desirability of European rabbits on American soil. Informed sportsmen have criticized the animal as a target on the grounds that it sits in its burrow all day or races into the subterranean maze at the first cry of a hound. Fearful farmers are fighting to outlaw the releases, pointing out that European rabbits thrive in other lands only by prostrating agriculture, preventing forests from reproducing themselves, and degrading pasture land into desert.

The American Rabbit Breeders Association is equally concerned. The association's officers know well that, while native cottontails and jack rabbits are relatively immune to myxomatosis, any successful colonization of America by European rabbits will be followed by introduction of the disease, endangering all domestic rabbits a mosquito can reach.

At a dog-racing track the greyhounds never seem to learn that the mechanical rabbit always wins the race. By failing to profit from past experiences, the human species may do no better in a race to keep European rabbits from conquering the countryside. From our modern vantage point, no one can predict whether the outcome will find rabbits or men in the lead. A new balance can be struck in either way.

> A ROADLESS *marsh is seemingly as worthless to the alpha-*
> *betical conservationist as an undrained one was to the*
> *empire-builders. Solitude, the one natural resource still un-*
> *dowered of alphabets, is so far recognized as valuable only*
> *by ornithologists and cranes.*
>   ALDO LEOPOLD: *A Sand County Almanac* (1949)

+o+o+o+o+o+o+o+o+o+o+o+o+o+o+o+o+o+o+o+o+o+o+o+o+o+o+o+o

CHAPTER · 14

Useful Wetlands

❀

As a land animal, one needing more and more living space, man tends to regard natural wetlands as his misfortune. Four fifths of the planet's surface is too deeply covered by oceans and lakes for any thought of filling them in. In consequence, the human species cherishes land that can be walked on. Indeed, the ancient Egyptians appear almost unique in happily anticipating the annual flooding of the Nile. They prepared for it, and counted on it to improve the fertility of their agricultural land.

Elsewhere man has tried to keep rivers in their channels at all seasons. He has drained swamps and marshes and pushed back the sea (as the Dutch have done for years), thereby increasing the area of dry land upon which food can grow and people live. Only in recent years and in a few parts of the world has this attitude toward wetlands begun to change.

The pioneers drained swamps by crude engineering methods. Agriculturalists of the nineteenth century had better tools and succeeded in eliminating at least 156,000 square miles of wetlands in the United States. Men enrolled in the Civilian Conservation Corps of the 1930's dug ditches, opened coastal marshes to the sea, and reduced the number of breeding places for mosquitoes. Despite all of these efforts, an inventory of wetlands in 1954 still showed some 116,000 square miles of the country where, for part of each year at least, water stands to a depth of from a few inches to ten feet.

These 116,000 square miles represent just over three per cent of the country. For Florida the percentage is forty-five, for Louisiana thirty-one. The state with the lowest percentage of all, surprisingly, is not Nevada or arid Arizona but moist West Virginia, whose mountainous terrain permits only .016 per cent of the state's area to be swamps and marshes.

The most continuous acreage of wetlands is in coastal marshes, but these areas are disappearing faster than any others. They are candidates for urban expansion, or for conversion into attractive shoreline property for recreational use, or for great airports, or for industrial land. At intermediate stages they become dumps for the refuse of big cities. All of these changes affect the wildlife of the land and also the fish at sea, if they need nursery shallows for reproduction.

We think of these changes as being caused by increases in human population. Yet the Chinese, so chronically overpopulated and deficient in usable land, appear to have been the first to find importance in wetlands. As far back as 3000 B.C. they were deliberately building fields to flood, there to raise rice, the staple food for half of the world's population.

Methods that originated long ago are still employed over much of the Orient. The fields are leveled, fertilized, and diked around. Then, when rain or hauled water floods the rice paddies, seeds are broadcasted or young seedlings are planted by hand. Peasants pull the weeds, giving the crop almost constant care. In this way the Orientals obtain yields of fifty to seventy bushels to the acre, far more food than comes from our best-managed wheatfields.

A wheatfield is expected to grow only wheat. Like an airport or industrial center, it is land on which a single use precludes all others. But a rice paddy produces fish as well. The warm shallow water becomes a rich culture medium for microscopic algae. These nourish both fish and small crustaceans. The minnows serve in preventing outbreaks of mosquitoes, whereas the small crustaceans and a few larger water plants are food for carp. The fish, moreover, aid in the nutrition of the rice, perhaps by adding ammonia to the water. As a result, the yield of rice in wet fields with fish runs from four to seven per cent higher than in those without.

Excessive numbers of minnows would interfere with production of carp. To prevent this, the Orientals ordinarily "weed" the fish population too, removing the small fish with fine-meshed seines and feeding them to ducks and pigs. In this way the wetlands aid still more in the local economy.

Carp tolerate the warm water, dodge the feet of peasants working on the rice fields, but otherwise lead such sluggish lives that they gain weight rapidly. In somewhat less than two months' time, the rice farmer can harvest from twenty-seven to forty-five pounds of carp to the acre.

The flesh of carp, although soft, contains proteins comparable in nutritive value to those in good beefsteak. Some vague appreciation of this must have antedated modern laboratory proof by better than forty centuries, leading to the introduction of carp and their pond culture in Europe prior to A.D. 600. The monks in monasteries of the Middle Ages are credited with developing several different true-breeding races of carp, each with special advantages. "Mirror carp" wear bright scales, making the fish decorative; goldfish are of this type. "Leather carp" have somewhat higher backs and fatter meat. These and other races reach maturity earlier than the type originally introduced, and seemingly produce more young. Carp culture came to England in the sixteenth century, but cold weather prevented its spread farther north.

In Europe the pond culture of carp continues in many areas. Often it is combined with production of wetland crops other than rice. Ducks and geese can dabble and cruise about, feeding on tadpoles, preventing excessive growth of reeds or pondweeds, and simultaneously fertilizing the pond —all without interfering with carp. Other fish, such as catfish, tench, and pike-perch, can thrive in the same shallow water so long as their numbers do not exceed ten per cent of the carp population. Medicinal leeches are raised as a by-product of carp ponds for a small but constant market. And if each pond is carefully enclosed, and if its fences surround plenty of vegetation, it can be used for additional gain from fur-producing muskrats and nutria.

In retrospect, it is easy to conclude that Spencer F. Baird, United States Commissioner of Fisheries in 1876, made a big mistake by introducing carp into America. At the time it would have been easier to list the desirable features of these fish than to predict the role they would take on this continent. A shipment of 345 carp came from Europe and were developed into a large breeding stock in ponds near Baltimore and Washington. The fish multiplied rapidly, and by 1880 some fifty thousand carp had been distributed. Within a few years hundreds of thousands were being shipped annually to all parts of the country.

Fishermen might never have complained about carp had European pond culture been imitated in this country. But neither Baird nor his assistant warned recipients of carp that the fish must be kept in captivity and fed when necessary with grains and kitchen scraps. Instead, the carp were freed in natural lakes and streams. This was the real mistake, for carp choose to root among the bottom vegetation, muddying the water if it can be muddied. Often they seriously reduce the depth to which sunlight can penetrate, and sharply decrease plant growth.

Now that the damage has been done, fisheries men are trying to measure it. To learn whether carp really destroyed the floating-leaf pondweed *Potomageton americanus*, a circular enclosure sixteen feet in diameter was built in one lake. Two months later the pondweed within the fence was found to be growing luxuriantly, whereas outside (where carp could reach the plants) the *Potomageton* remained in pathetic shape. At the same time, as much as nine inches of silt settled on the bottom within the enclosure, from material drifting in after being stirred up by the carp outside the fence.

At the western end of Lake Erie, cattails began to die off. Both carp and muskrats were suspected as the destructive agents, but whether they worked independently had not been learned. Wildlife managers M. L. Giltz and W. C. Myser tried to find the answer by fencing off an area with half-inch "hardware cloth," a screen through which neither carp nor muskrats could pass. Soon the men saw the plants within the fence growing taller, with more shoots from each buried horizontal stem, than in adjacent areas outside. Examination of cattails beyond the fence convinced the two scientists that carp undermined and exposed the horizontal stems in the muddy bottom. Then, as new shoots developed, muskrats could attack them easily. Working in series, the fish and mammals prevented any crop of fresh growth from reaching the surface of the lake and perpetuating the marsh.

Once carp are free in a lake, they resist control measures. Each five-pound female carp may lay about two hundred thousand eggs. In a year or two the few survivors from an extermination campaign can replenish the carp population as though nothing had happened. Often the cost in labor and materials is greater than the gain from removing carp. Occasionally, however, management pays handsome dividends.

A particularly dramatic change was brought about in Lake Mattamuskeet, seven miles north of Pamlico Sound in coastal North Carolina. There carp competed successfully with thousands upon thousands of migratory waterfowl. Prior to 1915 this lake was an open body of water with low marshland at the south end and one small outlet draining the entire seventy-eight square miles. In that year a stock company purchased the land and built what was then the largest pumping station in the world. Quickly the pumps drained the lake, allowing two fifths of the area to be reclaimed for rice culture.

When the company owning the property went into bankruptcy in 1932, the pumps stopped and the lake refilled. Two years later the federal government purchased the lake as a wintering ground for migratory waterfowl and built more canals to provide gravity drainage. In this way the Mattamuskeet National Wildlife Refuge came to contain some forty-seven square miles of open water and thirty-one of marsh and uplands. The average depth of the lake is but three feet, yet by 1935 the area had become a major fishing attraction due to tremendous populations of largemouth bass, crappie, and white perch.

Apparently the carp entered Lake Mattamuskeet from brackish coastal waters by way of the new canals. Slowly they replaced the sports fish and by 1949 had made the lake so turbid that submerged aquatic plants died off. At the same time the carp destroyed large quantities of marginal marsh vegetation whenever the lake rose after summer rains. Migratory waterfowl, for whose benefit the refuge had been set aside, no longer stopped in large numbers because they could find so little food at Mattamuskeet.

The refuge manager, Mr. Willie G. Cahoon, explained to us how he planned his countermeasures against the carp. He reasoned that if the adults could be prevented from coming into the lake from Pamlico Sound or from going from one part of the lake to the other (it is divided almost equally by an earthen causeway constructed in 1941–2), the number of spawning carp could be reduced. If large carp in the lake could be caught and removed, the program would leave chiefly young carp upon which predatory game fish could feed.

With a little encouragement from Mr. Cahoon, commercial fishermen nearby in North Carolina constructed over 150 carp ponds on private land. Then, under permit and close supervision, they used seines in special areas of Lake Mattamuskeet to obtain a stock of adult carp. Between the start of this program in 1945 and the end of July 1952, nearly nine hundred tons of live fish were removed, with a commercial value approaching $150,000.

At the same time Mr. Cahoon authorized construction of barricades in the form of weirs built of cypress slats, spaced to permit passage of slender white perch but to prevent the movement of fat-bodied adult carp between Pamlico Sound and Lake Mattamuskeet, or from one half of the lake to the other. Large gar, catfish, and bowfins were stopped too, but most game fish got through.

At the time these measures were introduced, the lake was so murky that shiny objects the size of a dime were invisible a mere six inches below the water surface. Seven years later we could see a silver coin of this diameter at a depth of three to four feet. Already several kinds of submerged vegetation useful to waterfowl had invaded more than twenty-three square miles of previously barren lake bottom. The surrounding marsh began to recover. The willows along the shores grew luxuriantly for the first time in memory, and annual bulldozing became necessary to keep them out of the lake itself. At an astonishing rate, the renewed availability of submerged plants led to an increase in visits by many kinds of waterfowl. Sports fishing, a benefit of secondary importance at Mattamuskeet, improved too.

Although smaller than thirteen other waterfowl refuges operated by the United States Fish and Wildlife Service, Mattamuskeet has become one of the most famous wintering areas for these birds along the Atlantic coast. Annually it attracts from 60,000 to 80,000 Canada geese, 80,000 to 150,000 ducks and teal, and hundreds of stately, whistling swans. One chilly November we climbed to the top of the tall chimney of the former pumping station and from the observation chamber looked out on this fascinating panorama of water birds. It seemed incredible that so few years had been needed for the change and that removal of one type of fish should have altered the habitat so tremendously.

Waterfowl refuges such as Mattamuskeet constitute about one third of the 150,000 square miles of refuge lands in the continental United States under the jurisdiction of the Fish and Wildlife Service. Probably they serve a fifth of the continent's waterfowl as feeding and resting places during migration, or as winter homes, or as breeding areas in summertime. To provide the food essential to wildlife using these refuges, more than a hundred square miles of croplands are either cultivated by refuge personnel or sharecropped, with virtually all of the government's share left in the fields for waterfowl.

The need for conservation of both wetlands and waterfowl came into prominence during the drought years of the early 1930's. Prairie potholes, ponds, and marshes had disappeared and waterfowl numbers reached their lowest ebb in

history. Yet the depression made it difficult to get funds with which to provide refuges, and many conservationists predicted early extinction of the nation's ducks and geese.

To meet this emergency, President Franklin D. Roosevelt scraped together eight and a half million dollars, most of it either from drought-relief funds or in the form of labor under the WPA program. The famous cartoonist Jay N. ("Ding") Darling became head of a special waterfowl committee. He, more than any other individual in conservation history, drew national attention to the plight of drought-stricken and excessively hunted ducks. He designed the first "Duck Stamp," and enlisted the co-operation of duck hunters, who henceforth would be required to pay extra for the bit of paper and affix it to their hunting licenses.

During the first year of its sale the Duck Stamp added almost six hundred thousand dollars to the funds available for waterfowl refuge work. In 1939 it hit the million-dollar mark. Ten years later the one-dollar price was doubled to meet higher costs, and in 1959 the hunter was asked to pay three dollars. Today the sale of Duck Stamps almost covers the budget of the nation's waterfowl refuges. And the annual harvest of waterfowl is believed to equal, if not exceed, the number taken at the time when market hunting of these birds reached its peak a century ago.

Many people have come to look upon the emblem of the wild goose flying and the sign NATIONAL WILDLIFE REFUGE as being synonymous with "sanctuary." They are only three-fourths correct, for, at the same time that Congress raised the price of a Duck Stamp from one to two dollars, it authorized the opening of not more than a quarter of certain refuge areas to public shooting, at the discretion of the Secretary of the Interior. At Mattamuskeet, on the one fifth open to public shooting in season, the hunters normally bag from thirty-five hundred to five thousand Canada geese and a like number of ducks, shooting them from blinds managed by the Conservation Department of the State of North Carolina. On adjacent farms, hunters account for at least as many additional birds.

The wildlife managers operate the waterfowl refuges for the benefit of the birds the sportsmen want. Water levels are raised and lowered, food is distributed, and maintenance is planned according to careful calculations. The aim is to raise the largest waterfowl population possible in America without at the same time endangering the crops of farmers. Other kinds of wildlife on refuge land prosper or decline according to how well their needs match those of the waterfowl and of the hunters whose Duck Stamps pay for the program.

On the migratory route followed by waterfowl paralleling the Pacific coast of America, the most spectacular national wildlife refuges are those in the basin of the Klamath River where it slants across the boundary from southern Oregon into northern California. Four refuges make up the cluster, each a multiple-use area with special emphasis on ducks and geese. Altogether, these refuges spread over more than 225 square miles, mostly in marsh and shallow lake with shores and islands bordered by the bulrush *Scirpus*, known locally as tule grass. Great swamps furnish food of importance to waterfowl.

For many years this region was a center for the market hunters. Today it is still a nursery from which come about 100,000 ducks annually. An additional 375,000 or more geese rest or stay through the winter, representing perhaps half of the goose migrants in the western flyway. A majority of the ducks from the entire northwest quarter of America use these refuges, settling for the cold months about halfway between the city of Klamath Falls, Oregon, and Sacramento, the capital of California.

Oldest and largest of the four areas is Lower Klamath National Wildlife Refuge, straddling the California boundary. Its 127 square miles were once the greatest production center for ducks, geese, and marsh birds in the western states. During World War I, however, Lower Klamath Lake in the refuge was largely drained in a misguided attempt to improve adjoining lands for agriculture. This action ruined the refuge for nesting waterfowl. Peat fires soon destroyed the exposed lake bottom. It burned to a depth of six feet or more, leaving only a vast alkaline, ashy desert, valueless for either livestock ranches or cultivation. In the 1950's a large part of the area was reflooded and restored to its original occupants, the migratory waterfowl.

Recently a new struggle for water has developed, pitting the federal refuge managers against state-appointed directors of the Tule Lake Irrigation District. The latter control pumping operations that move water from the drainage area of the Lost River in northern California, via a tunnel and canals, into Lower Klamath Lake. From there any excess drains off automatically into Klamath River and the Pacific Ocean. Tule Lake, in the middle of Tule Lake National Wildlife Refuge, is the natural sump of Lost River. By keeping the pumps working, the directors of the irrigation district can lower the water level in Tule Lake. In 1959, a water shortage year, the level was held more than six inches below normal, rendering huge areas of the refuge marsh inaccessible and valueless to waterfowl. Critics claimed that the water actually was thrown away in the hope that three or four square miles of flood-storage reserve land might be abandoned as refuge and opened to homesteading.

The implications of this struggle are wider than might be suspected from a short visit to the area. All four refuges in the Klamath basin are manipulated for purposes beyond providing sanctuary and controlled hunting in relation to the largest concentration of migratory waterfowl in America. By adjustments of water level and timely display of specially grown food, the immense flocks of birds are held on federal land until after the adjacent barley crops have been harvested and until the valuable rice in the Sacramento Valley to the south has been garnered. High water levels mean more ducks and geese on refuge property, less loss of young birds to predators, less mortality from avian botulism, and also significantly less depredation in croplands.

California's raisers of rice and barley, as well as those interested in hunting waterfowl, have never forgotten the 1930's, when drought in America's heartland led to crop failures and to disaster for ducks and geese through loss of wetlands suitable as breeding areas. These people included a

193

few who suspected that weather runs in cycles. If the critical years could be predicted well in advance, it might be possible to prepare for them and lessen their effect.

Weather is so variable that clues regarding long-term changes must be sought everywhere, not just in areas afflicted by recurrent droughts. Was it mere coincidence that when waterfowl from wetlands of the central states decreased so spectacularly, the formerly abundant American brant population shrank suddenly too? This little goose, whose head, neck, and breast are black, feeds along Atlantic shores where every tide brings another supply of water from the limitless seas. How could drought on land affect it? Yet between 1929 and 1932 the brant population fell to less than a fifth of former numbers. No longer could New England diners secure their favorite waterfowl in the quantities to which they had long been accustomed.

Those who knew the feeding habits of the brant supplied the immediate reason. The eelgrass, upon which brant feed, had died off in 1931 all along the Atlantic coast. Eelgrass is a strange, grass-like flowering plant of estuaries, bays, and sounds. Until 1930 it furnished about eighty per cent of the diet of brant. From 1931 on, only about ten per cent of the eelgrass remained. Great flocks of brant simply starved.

Fifteen years passed before the now eminent microbiologist Dr. Charles Renn contributed a real clue toward solving the mystery. He found that eelgrass had suffered a disastrous attack by *Labyrinthula*, a slime mold causing a "wasting disease." The strange feature of the epidemic was that *Labyrinthula* is a common infection of eelgrass, never before known to be dangerous. Perhaps a new strain had arisen—a more devastating one, behaving as a poorly adjusted parasite and killing its host.

Stranger still was the rapid spread of the epidemic to the British Isles and Europe, to the Mediterranean, and even to the coasts of Australia and Japan. Recovery in succeeding years proved slow and sporadic. Yet the situation could not be called unique, for records showed similar die-offs of eelgrass in the past—twice or thrice in a century. Apparently Julius Caesar merely had the good fortune to lead his armies into North Africa at a time when eelgrass was abundant in the Mediterranean and could be used as a hay substitute for his horses.

Without eelgrass, the fortunes of many creatures altered radically in the bays and estuaries. No longer did the long narrow leaves of the rooted plants provide areas upon which a thin film of microscopic vegetation could thrive, reproducing and supplying the needs of shellfish. Scallops and crabs, in particular, suffered from this change. So did the formerly plentiful little jellyfish *Gonionemus murbachii*, whose transparent, pulsing, one-inch bells had been a familiar sight in coastal waters. Small fish no longer had hiding places in the dangerous game of hide-and-seek with larger fish and squid. Fishermen discovered great schools of squid and numbers of slender pipefish—denizens of the eelgrass tangles exposed when this refuge died and disintegrated.

At Salcombe, on the coast of England, the extensive sandbanks that had been stabilized by eelgrass slowly eroded away. The whole shoreline changed, and inshore bottom materials were carried off by wave action to a depth of two feet or more. Buried stones came to light, and on them seaweeds commenced to grow. Animal life that had been common on the sandy bottom and among the eelgrass vanished or became quite rare. The whole fauna lost its former richness.

American coastal communities, which previously had ignored the warnings of microbiologists, suddenly needed no laboratory tests to demonstrate pollution from sewage. The untreated residues drifted back and forth with each tide, or washed ashore, now that almost no eelgrass remained as a restraining sieve loaded with decay bacteria ready to attack the sewage. And shifting sediments the eelgrass might have held now smothered clams and oysters, destroying a resource the shellfishermen had counted on. Sewage-treatment plants could end the pollution, but could not bring back the shellfish.

In only a few places has the eelgrass regained the full luxuriance of the late 1920's. The recovery of the brant has been even less complete, and the scallops, crabs, and mussels still have room for improvement. Yet fisheries scientists have noticed that restoration has been fastest from pockets of eelgrass that survived close to rivers, but in protected brackish bays that remained unscoured by spring floods.

Salinity alone could scarcely be the troublemaker, as eelgrass tolerates the full salinity of the sea and has spread sometimes to a depth of a hundred feet along the California coast. Nor does temperature alone seem important.

A new clue to the mystery was provided by occasional antics of the sea when a violent storm threw up a sand bar and isolated a lagoon, then later breached the barrier and permitted ocean water to flow freely into a lagoon pasture of eelgrass. This second change often seemed to trigger a disastrous outbreak of the slime mold *Labyrinthula*.

Salinity in an estuary, a bay, or a sound increases too when drought strikes the drainage basin emptying there. As feeder streams dry up, the river ceases to oppose the ebb and flow of tidal currents in its estuary. No longer does it dilute the sea significantly, and the environments in which eelgrass grows become more saline. Perhaps eelgrass loses its resistance to *Labyrinthula* under conditions of high salinity, particularly if the river does not flow and keep summer's heat from affecting the estuary. Or the slime mold may alter its habits, attacking with greater vigor when the salinity is high. Either of these explanations would relate the decline of eelgrass and coastal brant in the early 1930's to the nearly simultaneous reduction in waterfowl of the drought-stricken Great Plains.

Records of eelgrass die-off, twice or thrice a century, go back much further than reports on the state of the world's wetlands. Seemingly the two are related, and past history might be used to predict another round of widespread drought between 1963 and 1980. A hint may even foretell the change: series of unusually rainy years preceded the great droughts of the 1880's and 1930's. If this happens again, the wet seasons should be recognized for their true worth—an opportunity to fill the continent's reservoirs of wetlands against the desert conditions to come.

The pendulum of weather changes swings from drought to flood so slowly that we tend to forget that it will eventually swing back again. The few years at one extreme suffice to ruin farmers who have settled on land whose average supply of moisture is too scant. In other areas the rainy years cause such a spread of wetlands as to invite attention from drainage engineers. By leaving to wildlife the wettest and the driest territory, man has a better chance to realize his own hopes on the land in between.

THE earth holds a silver treasure, cupped between ocean bed and tenting sky. Forever the heavens spend it, in the showers that refresh our temperate lands, the torrents that sluice the tropics. . . . Yet none is lost; in vast convection our water is returned, from soil to sky, and sky to soil, and back again to fall as pure as blessing. There was never less; there could never be more. A mighty mercy on which life depends, for all its glittering shifts water is constant.

DONALD CULROSS PEATTIE AND NOEL
PEATTIE: *A Cup of Sky* (1950)

CHAPTER · 15

Waste Not, Want Not

❦

AIR, earth, fire, and water satisfied the Greek philosophers of Hippocrates' day as the elemental stuffs composing the universe. Hippocrates himself, the "Father of Medicine," insisted that man and all other living things were constituted differently, thereby having life. Their building blocks were also four: yellow bile, black bile, blood, and phlegm.

Today we can rewrite these ancient notions while supporting a modern view. The non-living background is seen as the atmosphere, the soil, the fresh waters, and the sea. These four sustain two great types of life—plants and animals—upon whose welfare man's future continues to depend. Even partial destruction of any among them worsens life for the human species.

For countless human generations the forests seemed endless. The prairies remained spacious enough for all of the useful livestock in the world. This was true while the early populations of mankind and domestic animals were so much smaller than today. But now the balance has tipped. No longer have we resources to waste.

Man wielded the ax and the saw so well that the remaining patches of virgin forest in the world can be counted. His plows ripped the native sod so thoroughly that in 1957 only a quarter of a square mile of unplowed tall-grass plains could be found in America. It is Tucker Prairie, just fifteen miles east of Columbia, Missouri, now set aside as a memorial tract. Already the scattered and seemingly vanishing prairie chickens have discovered the sanctuary. A few pairs of this short-tailed, hen-like bird are nesting there.

Now that more is being done to rescue from oblivion the remnants of once extensive natural resources, the interconnections between wastes and losses become more obvious. Wasted sawdust, for example, can mean lost shellfish. At Damariscotta, Maine, the bottom of the river is still covered to a depth of from two to three feet with sawdust, although the lumber-mill operators supposedly ceased dumping the material into the water half a century ago. An equal or longer time may have to pass before the wood fragments decompose and wash away, leaving the bottom clean and suitable again as a living place for clams and oysters, or as a spawning ground for coastal fish. That the river had a notable history before sawdust settled there is shown by the famous heap of oyster shells at the edge of town.

This object lesson has not sufficed in many quarters. Along various streams in New England, mill operators barely follow the letter of the law enacted to protect fresh waters from pollution with sawdust. True enough, the piles of waste are built on dry land, where only a few crumbs will blow into the adjacent river. But each pile can be rebuilt annually, as it is on an area the river will overflow in spring; flood waters regularly carry away the sawdust. The clever lumber merchant need not pay anything to dispose of his wastes. To do so would be merely to make money for shellfishermen or for those who would profit from summer tourists happily enjoying an unpolluted stream.

Today we talk and read about "one world" and wonder about ways in which entire nations can co-operate with mutual respect for the needs of each. In each nation, public-relations officers keep people informed about samples of national resources set aside for the enjoyment of all. But all too little has actually been done. The total area of park lands and sanctuaries preserved to perpetuate original America amount to barely more than half of one per cent of the country's surface. Many of these regions are spectacularly rocky or mountainous, or consist of deep-cut canyons. As such, they are little suited to agriculture, forestry, or any industrial use. Yet financial interests press constantly to obliterate these tags and pieces, either directly or through activities whose wastes would bring disaster.

The largest remaining stand of virgin hardwoods and hemlock in North America occupies three fourths of Michigan's Porcupine Mountain State Park, a tract of some ninety square miles located on the south shore of Lake Superior. It is also the largest roadless state park in the country, having been set aside in 1944 as a wilderness area—a living forest museum. Yet the state, unfortunately, owns only the "surface rights" in some sections of the park. Roots of the trees grow down into the property of copper-mining companies, the holders of "mineral rights." Recently the copper interests applied for additional privileges to improve their operations in mining, smelting, and reduction of the ore.

Few would question the rights of the mining companies to remove the ore under the park and extract the copper.

Although the ore is of low grade, its importance grows each year. Our needs for the reddish metal continue to rise slowly above the twenty-three pounds per person estimated in 1950, and, just as gradually, the average grade of ore sinks and extraction processes must be improved. In the eighteenth century no ore containing less than thirteen per cent of copper could be used. By 1900 the tolerable limit had shrunk to five per cent. In 1951 most of the copper refined throughout the world came from ore containing between .6 and .9 per cent of the metal.

Mere removal of the ore is unlikely to affect the topography of the Porcupine Mountains wilderness. For economical use of the ore, however, a smelter would have to be built near the mines, and hence close to Michigan's great park. The operation of smelters in America has invariably caused widespread damage to vegetation. Toxic sulphur-dioxide gas gets into the atmosphere despite continuing efforts to control or dissipate the fumes. The gas is poisonous to plants, and combines with rain and dew and mist to form sulphuric acid, burning virtually everything exposed to air.

To mining engineers, a roadless tract visited only by outdoorsmen who choose to canoe the streams or hike the trails may seem to be waste land, of negligible importance. But a person need have no financial stake in a piece of property to apply the word "waste" to land near a smelter. People who drive through the seventy-eight square miles of man-made desert at Ducktown in Polk County, Tennessee, will not forget how a copper smelter can affect the landscape. A more lifeless, baldly eroded area would be hard to find closer than the moon. Yet the eleven hundred inhabitants of Ducktown could visit similarly devastated land in Arizona, California, Montana, and in once forested parts of Washington state. These last were destroyed by fumes from smelters across the international boundary, at Trail, British

Columbia. The only comparable natural deserts we have met are downwind from sulphur-steaming volcanoes, such as some in Nicaragua.

Other wastes from a copper mine are highly poisonous to plants of almost every kind. The spoil banks of extracted ore heaped on land remain uncovered by vegetation until centuries of rain have leached away the toxic substances. During most of this time all drainage streams are completely uninhabitable to plants and wildlife, as well as unfit for any human use. Yet if the mining companies in Michigan follow a suggested alternative and dump the extracted ore in Lake Superior, the effect on aquatic creatures, water supplies, beaches, and shorelines can be equally disastrous.

At first glance, it seems incredible that a lake 350 miles long, 160 wide, and as much as 1,300 feet deep could be polluted enough to matter by any ordinary action of mankind. But pollutants dumped near the shore are kept concentrated at the very place where human interests begin. Even the great oceans suffer in the same way along the margins and over the surface wherever, through carelessness or callousness, wastes are allowed to spread. Dilution is too slow to reduce toxic materials to harmlessness before damage has been done. "Dilution is no solution to pollution."

Recently the Connecticut conservationist Dorothy Childs Hogner asked the Council of State Governments where she might find a river system in America sufficiently uncontaminated to be regarded as a model. Regretfully the officials advised her that they knew of no such ideal river in the country. All are polluted. Only remote streams retain the crystal purity that characterized all of the rivers of America three centuries ago.

Senator Karl E. Mundt of South Dakota has referred to pollution as a measure of civilization. We know how to avoid it, and cannot afford to continue it, yet too little is done in the right direction. The situation is worsening in many areas. The Missouri Conservation Commission found 900 miles of that state's streams polluted in 1949—about half of the total. A repetition of the survey in 1957 disclosed more than 1,200 miles of seriously contaminated streams, and a spread of the pollution from 69 to 74 of Missouri's 114 counties.

MISSOURI SCOREBOX—1957		
POLLUTANT	NUMBER OF STREAMS	NUMBER OF COUNTIES
coal-mine wastes	40	9
sewage	38	30
dairy wastes	23	18
petroleum oil	19	16
industrial chemicals	13	16
clay from barite washing	10	3
sawdust	7	7
lead-mine wastes	6	4
cannery wastes	3	2
all dangerous wastes		74

Water from coal mines, like drainage from coal wastes spread to the air, contains sulphur compounds derived from the slow decomposition of iron-sulphide (pyrites) granules. The acids formed from the sulphur compounds soon coagulate the mucus protecting the gills of fish, and the fish suffocate. The material frequently kills the food plants and animals upon which the fish depend. Usually it also renders the stream unfit as a water source for livestock or mankind.

Clay particles in sediments from washing barite ore (a

major source of the versatile element barium) render the water so turbid that sight-feeding fishes cannot find their prey. Month after month the pollutant can accumulate, blanketing the bottom, destroying the stream as a feeding ground or spawning area for many kinds of life.

The outstanding wildlife biologist Dr. Durward L. Allen has referred to the "deliberate expropriation [of rivers] as sewers." Sewage is like the wastes from dairies, canneries, and meat-packing plants in that it competes for oxygen in any stream. Gradually these organic substances are oxidized into inoffensive materials, but at the expense of wildlife. Usually the active sports fish are the first to suffocate as the available oxygen decreases. Shellfish may feed on the waste particles and survive, only to be condemned as human food because of contamination or off-flavor.

Harbors and other protected arms of the ocean are equally vulnerable to pollution. The narration of the General Electric Company's documentary film *Clean Waters* points out that the people of Los Angeles alone drive an unnecessary aggregate of fifty million miles a year to reach relatively unpolluted beaches and other aquatic recreational areas. In doing so they spend more than a million and a half dollars in gasoline, oil, tires, and worn-out cars in trying to avoid the consequences of pollution. At the same time their vehicles add significantly to the atmospheric pollution that has become so infamous in the Los Angeles area, as an eye-stinging, plant-burning, rubber-destroying smog.

In San Diego Bay each summer, according to the eminent microbiologist Dr. Claude E. ZoBell, as many as a hundred thousand people congregate on floating craft, ranging from small cruisers to large ships. Tides and currents cannot dissipate the sewage bacteria from these people fast enough to keep the population of contaminants from reaching fifty per cubic centimeter—more than one bacterium to each drop of bay water. The same sewage contaminants have been recovered alive as much as ten miles out to sea.

Natural populations of micro-organisms in the ocean occasionally reach two thousand times this density and liberate into the water various toxic materials with spectacular effects. Usually the chief offenders are minute armored cells, the dinoflagellates, seemingly on the border between the plant and animal kingdoms. Often their abundance is so great that they tint the water. Fishermen call the phenomenon "red tide," and know that the breaking waves free the minute particles as dust in such numbers as to cause sneezing and sore throats in coastal inhabitants. Customarily, a "red tide" is fatal to thousands of fish, shellfish, and even turtles.

The dinoflagellates of "red tides" are the pollutants, yet until the middle 1950's the reason for their sudden increase to such prodigious numbers remained unknown. Then a combination of factors fitted together, permitting accurate forecasts (but no control over recurrence). Water and air temperatures, river drainage patterns, and the amount of rain on land all enter the formula used to predict these calamities along the coast.

Clues to this mystery came from study of particularly destructive outbreaks along the coast of Peru, calamities that have come at long intervals but so often near Christmas as to have been dubbed *El Niño*—"the child." During a season of *El Niño* an abundance of rain falls along the otherwise parched coast, and the shore waters become warm instead of cold. Fish die by the millions, and the famous guanay birds, whose island nesting sites are mined for guano, are hard pressed to find enough food.

Ordinarily, the great eddying currents of the South Pacific drift northward past Peru and force an upwelling of cold waters rich in phosphates. These nutrients support a wealth of microscopic green plants in surface layers and, indirectly, a dense population of small animals. Young fish strain out the microscopic algae and capture the little crustaceans feasting there. Guanay birds and Peruvian fishermen take the fish, and perhaps enjoy the cooling effect of the upwelled water so close to the equator.

Occasionally some vagary of winds alters the flow of the Peruvian coastal current, and the upwelling movement is suspended. The nutrients diminish; the algae reproduce less vigorously; the crustaceans and other small animals become fewer; the fish tend to move to cooler depths; the guanays search ever farther for food; and the Peruvian fishermen hang up their nets, unable to catch enough to justify their time and effort. Sometimes a new current develops as a surface stream of warm water from the Gulf of Panama, passing southward along the Peruvian coast. From it the onshore winds pick up far more moisture than could be lifted from cold water, and rain begins to fall on land. Rivers that normally are dry for years on end suddenly gush seaward with great loads of sediment. *El Niño* is on its way.

The silty wastes discharged from the land sharply change the chemical nature of offshore waters and favor the reproduction of dinoflagellates. "Red tides" appear; poisons spread, and the oxygen content of the sea shrinks rapidly as the hordes of microscopic creatures use it in their respiration. Soon dying fish rise from the depths and are cast ashore by waves. Crabs drag themselves onto the land and succumb. The whole coast takes on a stench of death, with only the gulls and vultures feasting along the rain-swept beach.

Even where "red tides" never come, man is grateful when gulls and vultures act as unpaid servants in disposing of his garbage. They match human needs far better than the rats in the town dump, for the birds do not move into the house as rats are prone to do. Gulls and vultures scavenge at the community level, much as the domesticated dog did in the remote past when man was more solitary and a hunter. In recompense, gulls and vultures are protected in most parts of the world. In this fact can be seen a reflection of human readiness to take the expedient, cheap way of disposing of wastes—by dumping them on the land or in the sea.

In the North Temperate Zone, where garbage disposal has been organized most thoroughly, the herring gulls—our largest gray-backed gull with black-tipped wings and flesh-colored legs—have profited most. They have become expert followers of the shipping lanes, and gather in astonishing numbers as soon as the galley crew tosses the day's garbage over the stern. Herring gulls swarm over city dumps and find nesting sites in the cooler parts of the zone, usually favoring islands along the coast or in the larger lakes.

These and other gulls are quite willing to take living food if it is easy to seize. This was demonstrated when gulls from nesting sites on islands in Great Salt Lake descended at the critical moment in 1848 and saved the crops of the Mormon pioneers from destruction by short-winged, heavy-bodied "Mormon" crickets. The monument to the "seagulls" in the Temple grounds at Salt Lake City perpetuates the memory of this incident and serves as a strong argument there against any interference with reproduction of gulls.

California gulls, a smaller kind with greenish legs and paler mantle of gray on the back, thrive also in Utah. Those breeding on two different waterfowl refuge areas there increased by about 270 per cent between 1942 and 1950, then became somewhat stabilized at a population greater than ten thousand individuals. In 1957 wildlife manager Clyde R. Odin reported that gulls had stolen almost a fifth of the eggs from 317 nests of waterfowl he had under observation. He blamed the gulls also for a forty-per-cent loss of ducklings, and concluded that the supposed scavengers were really predatory—destroying at least thirty per cent of the water-fowl eggs and young on the refuge each June.

Dr. Arthur C. Twomey of the Carnegie Museum in Pittsburgh gave gulls an equally bad name for raiding the nests of waterfowl in the Deer Flat National Wildlife Refuge, near Boise, Idaho. Between six and seven thousand pairs of California gulls nest on a small island in the refuge. Dr. Twomey found that the parent gulls bring home and regurgitate intact for their young the eggs of cinnamon teal, eared grebe, black-necked stilt, American coot, and ring-necked pheasant. Ordinarily the young gulls peck at and eat the contents from these eggs. Bird-hunters, whose hunting taxes have gone to support the refuge, scarcely regard waterfowl eggs as garbage to be consumed by gulls of any kind.

Within the past fifteen years herring gulls have practically taken over Gardiners Island, at the east end of Long Island, New York. In 1940 no herring gulls nested there, and the island harbored the largest breeding colony of ospreys known anywhere in the world. Today Gardiners Island has more than fifteen thousand gulls. Instead of three hundred pairs of ospreys raising young on the island, only thirty-nine nests remain. Of these, just thirteen were occupied in 1958, and bird watchers fretted to see flocks of gulls following and diving at the parent ospreys whenever one returned with fish for the young.

Muskeget Island, halfway between Nantucket and Martha's Vineyard, Massachusetts, is now the largest center for herring gulls on the New England coast. At least eight thousand pairs nest there. As this number was reached, the populations of laughing gulls and terns, which formerly bred on Muskeget, shrank—chiefly as a decrease uncompensated by increase elsewhere.

New England lobstermen claim that gulls catch young

lobsters and are responsible for part of the gradual decline
of their business. All along the Atlantic coast, gulls do
search out the unarmored juveniles of the "living fossil"
known as the horseshoe crab. Older crabs, as much as fourteen inches across and two feet long, are safe enough even
in shallow water. But the young ones, while only two or
three inches wide, plow through the soft mud of tide flats,
feeding on algae and tube-building worms where gulls can
wade with ease. Man's encouragement of scavenging birds
may already have done more to hurry the horseshoe crab
toward extinction than any other environmental change
during the three hundred million years these strange animals
have lived in essentially their present form.

From the Netherlands to Sweden a change in point of
view has spread since the late 1930's. Gulls, particularly
herring gulls, still manage to find enough garbage so that
better than fifty per cent of their food is "civilized" items
such as discarded fish, meat scraps, and other wastes. To
many people, this free garbage disposal is too expensive, for
it is bought with the lives of other birds and even hares,

whose young the gulls pursue. Protection of the winged scavengers is being withdrawn, and wildlife officers are now systematically reducing gull populations.

These measures began in the Netherlands with the poisoning of some ten thousand herring gulls with strychnine. Further control was expected during World War II, when people began collecting herring-gull eggs to supplement scant supplies of more usual food. Yet at the end of hostilities about eighteen thousand nesting pairs remained in Holland. This was far above the ten thousand pairs tallied prior to 1930, although below the twenty-five thousand pairs in early 1939.

Removal of eggs merely leads a gull to lay another set. The process can be repeated over and over through the nesting season. To really reduce the number of nestlings, the eggs must be left in place but prevented from hatching. In Europe this is done by vigorous shaking, which kills the embryo inside. In America a poison spray is equally effective and less time-consuming. Following either treatment, the parent gull continues to incubate the eggs until the breeding season ends—with no progeny.

As man's attitude toward scavenging gulls changes, the damage they do to other sea birds may shrink. Yet a new danger now threatens the auks and murres and scoters and gannets that nest along northern coasts and dive for fish far out to sea. By the hundreds of thousands they are dying annually from another of man's wastes: oil on the surface of the ocean and its bays.

The oil hazard for sea birds began about thirty-five years ago when coal-burning ships were replaced by oil-burning and oil-carrying vessels. Modern ships follow a schedule that includes the expedient but wholly unnecessary practice of discharging engine wastes and flushing bunker tanks at regular intervals. Thousands of gallons of the oily discharge are left each year to float on the surface of the sea. Each gallon spreads within a week to cover better than an acre, and dooms any sea bird alighting there or surfacing in the oil film after a protracted dive.

Even a few drops of oil cause feathers to mat together, destroying their insulating value. In tropical waters this might be of less significance. But the patterns of ocean currents make fish at the surface available to sea birds chiefly in the bitterly cold waters near the great ship lanes between the Old World and the New. Any bird there which loses the insulation given by unfouled feathers soon dies of exposure. If the oil soaks into the plumage of wings and back, it interferes with the bird's ability to fly and dive, and condemns it to slow starvation. According to Mr. Leslie M. Tuck of the Canadian Wildlife Service, a spot of oiled feathers on the belly, no larger than a twenty-five-cent piece, will kill a murre. The survivors of oil fouling are likely to be those few sea birds whose encounter with the floating menace occurs just prior to the annual molt and so close to shore that they can pull themselves out of the water.

In the western Atlantic the murres bear the brunt of man's careless dumping of oil. These small relatives of the penguins can fly if they have to, but spend much of their lives drifting with the currents, fishing on the surface of the ocean. As arctic ice packs drift southward early in the year, the birds become concentrated along the south and southeast coasts of Newfoundland, near the shipping lanes and oil. Eider ducks, which congregate in bays and estuaries close to land during the winter, become fouled by oil slicks and by the sticky, tar-like blobs produced on the surface after waves and sun have modified the floating residues. Yet only the few hardy people who go swimming in these cold waters become acquainted, through black-stained feet and bathing suits, with the actual material causing such destruction of wildlife. Farther south, the same pollutants earn the ire of more bathers, but cause less damage to birds because there are fewer waterfowl in warmer waters to begin with.

Riding the ocean currents, the oil travels for hundreds of miles, extending the damage done by occasional disasters such as the sinking of an oil tanker. A German bird-watching society at Wilhelmshaven reported 275,170 sea birds dead from oil fouling along the coast of the North Sea in 1955, presumably due in large part to the breaking up of a Danish tanker in a storm. Pollution of the Baltic at various times in recent years has all but exterminated the once common old-squaw ducks there.

Whether a gradual change from petroleum products to nuclear power will mean the salvation of the sea birds is still in doubt. We know how to dispose of waste oil in port, and how to separate it, rendering it harmless, at sea. Few ships, however, carry the necessary equipment. As Dr. S. C. Martin of the U.S. Public Health Service commented before an international symposium on water-pollution recently, "It is simpler and cheaper to use the public property for their waste than to do the thing right."

Radioactivity is neither simple nor cheap. We have learned how to release a powerful demon from its atomic box, but lack knowledge of how to order it back in. We can turn on artificial radioactivity, yet must still wait for decades or millennia until nature turns it off again.

Even our earliest tests of nuclear power are producing wastes no one knows how to handle, and our use of it remains limited largely by the problem of where to store radioactive residues until they become harmless to mankind.

Scientists continue to follow through the South Pacific the slowly drifting masses of contaminated water from each test explosion. They seek to learn what happens to the dangerous compounds. Microscopic plants, drifting in the water, provide a sort of sponge for radioactivity, but they pass the hazard along to the small animals eating the algae. Shellfish strain out the minute particles and store dangerous cesium-137 in the very muscle tissues man most enjoys. Fish eat the plankton crustaceans as well as the algae and may accumulate perilous strontium-90 in their bones.

No international agreement has been reached on how much radioactivity can safely be dumped into the ocean. Conservative methods of handling these unfamiliar materials cost tremendous sums, ones only a rich country can afford. The same cannot be said, however, for most other pollutants still being freed in bodies of water or in the air. The cost of treatment that renders them harmless is not so great. The cost of failing to do so, in terms of food for mankind and of wildlife resources, is already incalculably high, and is increasing every year. So are the human population and the quantity of wastes to be disposed of. The wants of man and other living things multiply—for more fresh water, more unpolluted sea and air.

To most people, as to those who write our dictionaries, a forest becomes merely "a tract of land covered with a dense growth of trees." These people would define a city as "an area occupied by closely spaced buildings," and ignore its men and women, its rats and mice, and its ailanthus trees.

A forest is more than an area covered by trees. In many ways, it is like a city—nature's city—constructed and peopled with trees, birds, insects, shrubs, mammals, herbs, snails, ferns, spiders, fungi, mosses, mites, bacteria, and a myriad of other living forms.

JACK McCORMICK: *The Living Forest* (1959)

CHAPTER · 16

Tree Country

THE world's oldest living things are trees. Along the crest of the White Mountains in the Inyo National Forest, California, more than a hundred gnarled pines on exposed slopes are known to exceed four thousand years in age. These trees are still growing, still producing seeds, still potential parents. A mere 250 generations of such patriarchs takes us back a million years to the beginning of the Ice Ages and a time when primitive man had yet to learn how to make fire. A comparable number of human generations have passed since the early Britons piled one stone upon another in building the great temple at Stonehenge.

Among earth's largest living things are coastal redwoods, at least one of which stands 364 feet high, with a diameter of more than twenty-two feet at the height of a human chest. Several redwoods have been found to be more than three thousand years old, yet still very much alive. These trees seem immune to fungi and insects, even to being struck repeatedly by lightning. Only one redwood giant, in fact, has been known to die a natural death. In a protected grove it toppled to the ground after several weeks of unusually rainy weather and turned the soil around its roots to fluid mud incapable of resisting the leverage of storm winds against the tree.

Three thousand years ago, when some of today's redwoods were seedlings, Assyria and Babylonia had separate empires. While much has happened in human affairs and floods and earthquakes, landslides and volcanic eruptions have altered the face of the planet, the bristlecone pines and the redwoods remain almost unchanged. Do they live on and on until the earth no longer supports their roots or until the weather takes too many turns for the worse?

Before man began using tools, the world's forests spread over many areas where no trees grow today. Then the twin destroyers—fungi and insects—and fires of natural origin, such as those started by lightning, were "nature's lumbermen." Probably their effectiveness has remained fairly uniform over a very long time, the combined toll averaging at least a quarter of the total annual growth achieved by the forests.

Today foresters deplore losses to insects, fungi, and fire because wood has become man's largest crop. Over most of the world, trees are being cut more rapidly than trees of the same kinds can regenerate. In many areas this procedure is deliberate: it exposes more soil for agriculture. In others it is merely a measure of the degree to which demands for wood exceed the productivity of the woodlands.

Present knowledge of what forest resources remain is based largely upon a survey made in 1948 by the Food and Agriculture Organization of the United Nations. Thirty per cent of the world's land surface was reported to be forested. On a per-capita basis, this would now average less than four acres per person.

Most of us expect a forest to yield products for which there is a national or world demand, products such as sawed lumber, pulpwood, ties, or poles. Yet large tracts reported to the United Nations as forests fail to meet this expectation, for they bear only thin stands of scrubby vegetation and furnish merely fuel for local people who gather short sticks by hand. The United States often lists among its "unproductive forest" great tracts from which all usable trees were cleared before the land was deeded to the Forest Service. These areas are physically capable of producing crops of wood. Under proper management, they can be expected to do so within a century of two. African and New Zealand forests, on the other hand, include immense territories clad in thorny shrubs. Fully sixty-four per cent of Africa's "forests" are *bushveld* of this character. The same is true of seventy-two per cent of forest land in New Zealand, and of thirty-four per cent of forests in the world as a whole. Productive forests on our planet total only about two and a half acres per person.

To be useful, a forest must also be accessible. Generally this means that the forest products must be valuable enough to make the building of good access roads worth while. Unfortunately, swamps and mountainous terrain make many productive forests inaccessible, and at present no use is made of them. The amount of woodland from which products are available is thus reduced still further—to less than an acre and a half per person.

These figures are more meaningful when compared with use of wood in America. Each person in the United

States uses up or stores away annually in paper products—books, cartons, newspapers, tissues—about the amount of pulpwood growing under average conditions in an acre and a half of trees. Yet paper products are merely our third-ranking use for wood. We use slightly more for fuel, and over five times as much for lumber from saw logs. Altogether, the average United States citizen requires the products from about thirteen acres of woodland annually—eleven acres just to supply the three needs mentioned. Citizens of other countries, who would like to use wood products as freely, have a right to inquire where in the world the trees can be found to satisfy their wishes.

Even when only the productive and accessible forests are considered and a total of ten per cent of the world's land surface is seen to be forested with trees that can be used commercially, the answer is not complete. People are highly selective in using trees. Paper, upon which publishing and education are so dependent, can be made from a variety of different woods. But in building, industry, and agriculture the choice is usually pine or some other softwood (conifer). Hardwoods, which come from the broad-leaved flowering trees such as oak, beech, maple, balsa, and mahogany, tend to be used chiefly for specialty products.

Today nearly a third of the world's accessible softwood forests (and a fourth of the inaccessible softwoods as well) grow in the United States and Canada. The only other really substantial softwood forests are in northern Europe and the U.S.S.R. Of these countries, Canada alone exports more softwood than it imports. In spite of all the lumber and pulpwood we purchase from outside the country for domestic use, we are cutting timber from our fifty-five thousand square miles of accessible softwood forests almost twenty-four per cent faster than it is being replaced by regeneration and reforestation. For trees of saw-timber size, the excess of drain over growth is closer to fifty per cent. At the same time our fifty-two thousand square miles of accessible hardwood forests are growing eighteen per cent faster than we are using the wood from them.

It is hard to realize that the enormous tropical forests in Brazil and the jungles of the Congo are between ninety-seven and ninety-eight per cent hardwoods, chiefly of kinds no one wants for anything more basic than an occasional chair, or a grand piano, or decking on a sporty sailboat. Sweden has less than a seventeenth as much forest area as Brazil, but in terms of softwoods Sweden is three times as rich.

Any virgin forest, whether of coniferous or hardwood trees, can be awe-inspiring even to the professional naturalist. Charles Darwin recalled such emotions when, in his *Journal of Researches* from the voyage of H.M.S. *Beagle,* he wrote of Brazil in retrospect:

> Among the scenes which are deeply impressed on my mind, none exceed in sublimity the primeval forests undefaced by the hand of man. . . . No one can stand in these solitudes unmoved.

Yet the word "solitudes" is too easily assumed to be mere poetry. Wherever the trees grow densely and a majority of them are mature, almost no other kinds of life can be found. Darwin himself remarked on this aspect of the impenetrable kauri forests in New Zealand: "In the woods I saw very few birds."

That the "forest primeval" is practically a biological desert was scarcely realized until Herbert L. Stoddard studied bobwhite quail in Georgia's woodlands between March 17, 1924, and June 30, 1929. From his observations has come the idea of "edge" as a home for quail and many other types of wildlife, whether that "edge" is the forest's border or the many openings and clearings within it.

When the pioneers from Europe began opening up

America's woodlands, they provided more and more edge. Soon they concluded that the edible mammals and birds seen frolicking along the boundaries and among the second-growth saplings had previously been denizens of the forest itself. Actually, these animals were a crop, raised inadvertently—creatures profiting from the freshly available seeds and foliage of quick-growing vegetation where formerly only giant trees had stood.

America's great modern heritage of national forests began when any timber-bearing area was expected to yield only trees. The Fifty-second Congress granted broad powers to President Benjamin Harrison in 1891 to save forest resources for the future, as rapid exploitation threatened to destroy them completely. The first reserve later became the Shoshone National Forest along the Continental Divide in the Wind River Mountains.

The U.S. Forest Service, a division of the Department of Agriculture, now looks after more than 293,000 square miles—about a twelfth of the total continental area of the country—as a heritage aggregating somewhat more land than the whole of Texas. Trees are its major concern, but multiple use has become a goal. America's dissected forests are seen to be important also as conservators of water and wildlife, and as havens for vacationers who wish to camp far from the crowd and the urban world; in some areas a limited amount of controlled grazing by cattle and sheep is permitted. This undertaking is operated on behalf of the public under a directive set forth by James Wilson, Secretary of Agriculture in the cabinets of William McKinley, Theodore Roosevelt, and William Howard Taft: "for the greatest good of the greatest number in the long run."

The national forests constitute the largest area of timberland under single management. Not only are they one of the few federal enterprises to return a profit (about three dollars for every dollar spent), but they serve also as testing grounds for ideas as to how greater good can be produced "in the long run." Foresters need to know which method of harvesting the trees will best promote natural regeneration of the forest. Which will encourage wildlife without leading to disastrous overpopulation? How can recreation values and watershed protection be combined?

Increasingly, the discoveries made on federal timberlands are being applied also in "industry forests"—those belonging to companies whose prime concern is paper, pulp, or saw timber, and to private enterprises for whom trees are secondary to mines, oil wells, hydroelectric power, railroads, or salable water. Some 300-odd companies manage nearly as much timberland as can be found in all of the country's national parks and state- and community-owned forests put together. Rarely are their holdings smaller than a hundred square miles apiece, and the average is closer to twice this size.

Almost all the rest of the commercial forest in the United States is in blocks less than a square mile in size. Better than four and a half million people own it, their tracts averaging only one thirteenth of a square mile each —forty-nine acres. Yet ninety per cent of the wood products reaching the market annually come from these small tracts, and two thirds of this wood is removed by management practices the Forest Service deplores as "poor or destructive." The tradition of free enterprise is too strong in America. Both state and federal governments are still reluctant to step in and prevent malpractice, even where removal of trees is done in a fashion likely to harm someone else's land by causing loss of topsoil, silting of streams, spring floods, summer drought, crop failures, reduction of water supplies for valley towns, elimination of recreation spots and of wildlife.

Pessimists have picked 1970 as the year in which a continuation of present mismanagement in the small wood lots will bring an effective end to America's timber supply at prices most of us can afford. To prove the pessimists wrong, American Forest Products Industries, Inc., is trying valiantly to encourage deliberate "tree farming." Eventual yields of twenty-five dollars per acre per year from wood products are expected, which is more than the return from most other uses of the same land.

To qualify as a "tree farm," the property must be

> privately owned, tax-paying and dedicated by its owner to growing and harvesting repeated crops of timber. The forest must be managed in accord with good practices. Protection against fires, insects, disease and destructive grazing must be in basic management programs. . . . Fire lanes and access roads both can provide forest "edge" so important to wildlife. . . . Furthermore, each harvest must provide for establishing the next crop of trees.

Most of the larger holdings are accredited tree farms, and thousands (but not yet the needed millions) of smaller ones also are enrolled in this nationwide tree-growing movement.

Good forest management coincides with protection of watersheds, soil conservation, maintenance of recreational values, and improvement of conditions for many kinds of wildlife. To further the raising of trees and game as twin crops, the eleven-per-cent federal tax on sporting arms and ammunition has been put to work since 1938 in most states under the Federal Aid in Wildlife (Pittman-Robertson) Act, supervised by the U.S. Fish and Wildlife Service. Each project is designed to yield basic information on wildlife relationships. One discovery after another has come from these sponsored researches.

We all know that squirrels plant acorns and other seeds that can grow into trees. Yet until the habits of squirrels were investigated, no one was sure how many of the cached morsels were dug up again and eaten. Now it is clear that the critical factor is whether the seed has germinated or not. A squirrel will not eat a sprouted acorn. If a tree farmer wants to raise trees and provide squirrel-hunting at the same time, he can plant sprouted seeds and be confident that the squirrels will not molest them. This method not only allows the new tree to rise where it will be spaced properly in relation to its neighbors, but also permits use of a new scientific development—breeding better trees by using selected seed, taking advantage of fresh information on the genetic strains of long-lived plants most suited to soil, weather, and specific uses. Inheritance in trees has become a new scientific frontier.

Often the timber-raiser is tempted to specialize. For the benefit of his trees, he may want to apply insecticides and fungicides by aerial spraying, and shut his eyes to the damage produced among useful forest insects and wildlife. Only through research can anyone predict whether gains will outweigh losses. Hardwood trees have flowers and often benefit from cross-pollination by insects, such as bees. The extent of the gain has been measured for oaks. Premature windfall of acorns is only 17.4 per cent if they are cross-pollinated, 61.5 per cent if self-pollinated because no insects arrived. Acorns from cross-pollination are 46.6 per cent heavier, and the seedlings from them sturdier than those from self-pollinated seeds.

In a forest operated for multiple uses, the economic results sometimes are surprising. Red spruce and white spruce from the northeastern United States and adjacent Canada are still the favorite sources of pulpwood for paper products. Spruce and moose go together so regularly across the boreal life zone of America that this broad evergreen sash hung diagonally from Alaska to New England is often referred to as the "spruce-moose formation."

During the 1957 season in Ontario, some 19,262 moose-hunters in the spruce zone paid $364,883 into the provincial treasury for the right to hunt and kill one moose each. These license fees gave the Ontario government far more income than it could have collected in taxes from the sale of spruce produced annually on the same area of land. The $364,883 exceeded even the sale value of 110,000 cords of pulpwood.

The moose-hunters spent a total of $2,815,000 that year on their sport—an average of $147 each. Nearly six thousand of them succeeded in getting a moose apiece. They harvested about six per cent of the moose herd, and could have taken three times as many without harming it. Presumably, if three times as many moose could be harvested annually from the area, three times as many moose-hunters would be willing to spend $147 apiece on the chance that they would be successful. This amount of money is economically equivalent to the sale of a cut totaling 400,000 cords of spruce pulpwood, and the license fees would bring the Ontario government more than a million dollars.

Almost all of this moose-hunting, however, was on private property, within a mile of some road—usually a road built by a pulpwood company. The question was raised: should the Ontario government not build access roads for moose-hunting, as a means of increasing revenues from license fees? This public expenditure would save the pulpwood-growers thousands of dollars in road construction and provide them with a gain from letting hunters shoot moose raised on tax-paying timberland.

From our own travels in America and Africa we have come to realize that the United States and Canada are uniquely favored for management of forest lands on a multiple-use basis. Much of the money spent by hunters and vacationers goes to local residents who run gasoline service stations, groceries, taverns, motels, and taxidermy shops, and to guides and renters of boats. Often these people earn enough during vacation and hunting seasons to be able to live comfortably the rest of the year on soil that would yield no adequate return from agriculture. The local folk have a vested interest in the perpetuation of our forests and wildlife even if they do not own them.

In other parts of the world these conditions have no counterpart. The public lands and large private estates to which vacationers go to hunt wildlife with camera or gun in Africa are hosts to fully equipped expeditions fitted out in distant cities. The "white hunter" brings the complete safari, providing comparative luxury for his paying guests. Well-trained black men from the city come along to set up camp, prepare the drinks and meals and beds, and help guide the visitors to the game animals. Then the whole party disappears again, having spent no money in the vicinity. Usually the local African residents remain at a distance, with no financial gain from seeing the safari come and go. It is no wonder the black man in Africa regards game preserves as only for white visitors. Is the local population to profit only by the age-old practice of poaching for meat and horns? Or by cutting the thorn trees as fuel and using the land as grazing territory for poor herds of domestic animals?

We discussed this with the directors of wildlife conservation in Africa, and suggested that when the herds in sanctuaries required culling to match the carrying capacity of the land, the meat might be given to adjacent native tribes, thereby impressing the people with the very real dividends to be expected annually from a game reserve. Surely the excess wart hogs and various kinds of buck would be acceptable; excess animals need not be left for the vultures. Could the wildlife not be regarded as public assets that match the economy of the soil better than domestic stock and yet yield food for primitive Africans?

The officials assured us that such a program had been proposed several times, recommending specifically that the meat be donated through the local tribal chiefs. In this way the game warden could reward those chiefs who helped control poaching along the sanctuary borders. But the federal government vetoed the plan, with the statement that wildlife conservationists had no right to concern themselves with native affairs beyond prosecuting poachers caught on game reserves.

At many points in Africa south of the equator, tree farms of a sort are rising at an amazing pace. Few of the indigenous hardwood forests remain, and valuable trees grow too slowly for replacement to be economical. A magnificent stinkwood tree may need two hundred years to reach full size. Bare hills are being planted to fast-growing eucalyptus from Australia and pines from the Mediterranean and America. In the eastern Transvaal province of the Union of South Africa and in the British protectorate of Swaziland we saw that these new forests are in small blocks separated by broad fire lanes. The Cape Province of the Union raises forests without treeless corridors because rain falls so frequently that the forest duff never dries out and fire is not a hazard. Foresters assured us that when the trees mature, a limited number of people from the vicinity will gain through employment as wood-cutters.

To a person familiar with lumbering operations in America or northern Europe, these African undertakings are astonishingly simple. Usually the still slender trees are felled by two men with a crosscut saw, then separated by one man with a bow saw into lengths of from six to twelve feet. The length depends upon the amount of curvature in each trunk. These short pieces are snaked out by one-mule power, commonly to be stacked by men beside the road where a truck can take them away. Generally the yields are too small to pay for use of chain saws or larger power tools. No advantage accrues from time-saving equipment, for the problems of maintenance mount where mechanics are few

and spare parts must be brought from far away. What, we were asked, would the wood-cutters do with the time saved? The need is for more work for more people.

Poaching of trees in these African afforestation projects is negligible, even where the native people on adjacent mountain slopes cook their food over dry cow dung because they can gather no fuel wood. To these natives, neither a eucalyptus nor a pine resembles anything useful in their traditional way of life. The branches are not slender or strong enough to make wicker reinforcement for mud huts. The foliage is unsuitable for thatch. Wooden beams have no place in this architecture, nor could the blazing fire from short lengths of tree trunk be accommodated inside the hut, where the fire belongs.

How much wildlife will find a haven in a pure stand of pine or eucalyptus remains to be seen. Even where the two types of trees are mixed, the forest remains open at the bottom like a park. Almost no other plants grow on the woodland floor. In most of the afforestation areas the indigenous vegetation was eliminated years ago by chronically starving cattle, sheep, and goats. Unless a new flora is introduced and some game animals as well, the chances for colonization are slim. Nor are birds likely to carry the seeds, as they would in a corresponding region in Anglo-America. Birds settle at their peril near the grazing grounds of Africa, for sharp-eyed children guarding the livestock are always ready with a stone. Half-starved dogs run down any game animal they can detect.

The native African is rarely sophisticated enough to plan for the future. The future should take care of itself. If the sole remaining tree in a village is cut into firewood and no replacement is planted, only the white man is likely to inquire what shade the children will sit under when they are grown men debating the affairs of the community. The spread of cement and tarmac at the expense of plants in a modern city has its smooth counterpart in the bare earth around the huts of indigenous people.

In most parts of the world, primitive man has been willing, even eager, to adopt the external signs of civilization. But often, in relinquishing the stabilizing, stylized tribal ways, the native finds no satisfactory substitute in white man's methods. They change too often! If the white man has tea one morning, why does he want coffee the next? His conservation procedures have no more regularity.

In Africa the European settlers cleared the trees from one area to make land available for efficient agriculture. On an adjacent area the natives were stopped from clearing the trees and imitating the white man, because doing so would destroy the habitat for wildlife. The difference between the soils of the two tracts made no impression. Moreover, the European sallied into the bush country whenever he felt like it and shot meat animals. Sometimes he gave his "boys" the meat and kept only the head for a trophy. Yet the same European objected when native people used their traditional weapons to get the same meat when they were hungry. Nor could he teach the natives his dislike of lingering death, or that rapid destruction of wildlife would follow too persistent hunting.

The Negroid inhabitants of British East Africa, the Central African Federation, and the Portuguese colonies of Angola and Mozambique had barely become used to one set of rules before the white man changed them. Suddenly he saw the necessity to halt a threatened southward spread of the contagious virus disease called cattle plague (rinderpest). He began hiring thousands of natives to cut the bush and stack it in great fences from east to west. The first one, twenty-six miles long, was finished in as many days. The next, along the Northern Rhodesia–Tanganyika border, soon stretched for eighty-six miles. Each barrier had an average height of eight feet, and was never less than seven. Being thick and opaque, it discouraged the smaller animals from crawling through and the larger ones from jumping blindly to the opposite side. Only the elephants passed. Calmly they unstacked the top layers and stepped over the remainder; native labor had to restack the fences several times a week. Elephants form no reservoir of cattle plague, and the fences successfully barred infected wildlife from carrying the contagion into grazing lands beyond.

Once rinderpest ceased to be troublesome, the Europeans put their minds toward ridding Africa of the blood parasite *Trypanosoma* carried by the blood-sucking tsetse fly. Indigenous animals seem immune, but domestic livestock develop fatal nagana disease, and man comes down with equally disastrous sleeping sickness (trypanosomiasis).

The many parts of Africa closed to cattle and dangerous to man because of trypanosomes could be opened if only the tsetse flies were eliminated. Entomologists reported that tsetses died if kept in the sun. Ordinarily the flies rest on trees and shrubbery between one blood meal and the next. If great areas of Africa could be shade-free, surely the trypanosome problems would end.

Native laborers were put to work eliminating everything higher than short grass from a broad band across Africa. Without understanding why they were being ordered to cut and burn the firewood so uselessly, they worked at top speed. Meanwhile the entomologists continued to watch tsetse flies. Near Lake Victoria a census was taken of the two major kinds found locally: *Glossina palpalis* and *Glossina pallidipes*. Around Nzalagobi the population of the latter outnumbered the former ten to one, hence seemed more menacing and received special study. Each fly, on the average, appeared to feed every fourth day, and to withdraw about sixty milligrams of blood. For about three months of the year (the "Fly Season") this regimen continued, giving a tsetse fly some forty-three meals, of which fifteen seemed to be on bush pig, thirteen on bushbuck, and the remaining fifteen on a wide variety of other wildlife. Each host animal seemed to support an average of 1,163 flies, being bitten by two insects in each five-minute period from dawn until sundown. This took from each pig or antelope about eighteen cubic centimeters of blood daily, just to feed tsetse flies.

Recollection of these figures passed through our minds whenever a tsetse fly darted through the open window of our car in the thorn forests of central Africa. Despite their excellent vision, however, the insects proved easier to catch than a honeybee as they buzzed bee-like against the rear glass pane. Tsetses ordinarily fly under a four-legged animal and settle on its belly. The Harris trap, devised in Natal's rhinoceros country, takes advantage of this habit by providing tsetses with imitation belly surfaces. And when we were stopped to be "deflyed" on our way toward the main coastal road from the Charters Creek Wildlife Reserve of Natal, it was the underside of the car that the efficient team of men sprayed most thoroughly in the closed, drive-through treatment shed.

Constant guard is necessary to eliminate shade in which tsetse flies will settle. In the rainy season, plants grow so rapidly that maintaining a shadeless zone across Africa proved almost impossible. As soon as growth appeared from cut stumps, the tsetses moved back again. In Mozambique the government looked into the possibility of killing the stumps with chemicals, letting them dry for a few ·months, and then burning them—a real scorched-earth policy. Many people wondered whether even this would help. Others questioned whether it was economically feasible.

Perhaps it would be easier to eliminate the game animals. Over a period of five years this too was tried, in six hundred square miles of Tanganyika and in various northern parts of the Union of South Africa, although not in famed Kruger National Park. Every bush pig, wart hog, zebra, elephant, and cud-chewing creature from the stateliest giraffe to the smallest duiker was shot. The flies were expected to starve, become extinct, and let man move into the regenerated thorn forest with his domestic animals, fearing no trypanosomes. Joyfully the native people helped with the destruction. So much meat had never been available to them before. But the tsetse flies did not vanish entirely. Even snakes and crocodiles seemed to harbor trypanosomes and serve as hosts to the flies.

The government of Southern Rhodesia sponsored a program of intensive and sustained hunting between a pair of game-proof fences ten miles apart, running across the country from east to west. The ends of this barrier lay in fly-free territory, and parasitologists believed that no trypanosome-carrying tsetses could enter the sterilized belt from farther south. As soon as all game had been destroyed in the initial belt, the parallel fences were moved north together and the process was repeated. Theoretically, new populations of trypanosome-free game would spread into the cleared regions.

By 1945 about ten thousand square miles supposedly had achieved trypanosome-free status. In 1945 alone, official hunters killed 7,518 dainty little duikers, 3,242 handsome steinboks, 912 of the cliff-climbing klipspringers, and 187 tawny oribi antelopes without finding in any one of them a trace of trypanosomes. Comparable programs of extermination flared in the Union of South Africa, Northern Rhodesia, Kenya, Tanganyika, and the Sudan.

Such wholesale slaughter offended many people, although few of them had black skins. Mr. J. Stevenson-Hamilton, retiring after many years as warden of Kruger National Park, reflected sadly on the whole sorry affair. He agreed that tsetse flies decreased in numbers wherever all, or nearly all, of the larger wild animals were killed. But he insisted that a residue of the blood-sucking insects remained, surviving at the expense of sleeping owls, squirrels, and some of the smaller buck that escaped elimination. For years the presence of flies might go unnoticed. Yet as soon as domestic animals were introduced, the tsetses would surely flare up again. Mr. Stevenson-Hamilton urged officials to follow the policy found so satisfactory in Kruger Park: let nature go, with as little human interference as possible.

Between 1948 and 1949 the rules were changed again, to the consternation of native people who had grown accustomed to meat from the slaughtered wildlife. No longer was bush to be cut, making firewood abundantly available. No game animals were to be killed, even as food for a healthy, rapidly expanding human population. Instead, airplanes would now fly low over the scrubby vegetation, spraying it at frequent intervals with mists of DDT. Surely this would eliminate the tsetse fly and other insects too. But what is gained by saving cattle from nagana only to have them starve for lack of pasture? Or for a man to be safe from sleeping sickness when wood to keep him warm and meat to nourish him are locked up by arbitrary orders? Agitators found many who agreed that Africa was happier before the Europeans came.

The eminent conservationist Dr. Paul B. Sears sees in the Western Hemisphere a similar situation, one in which the old ways of the indigenous people matched the needs of the land far better than the ideas of incoming conquerors. For untold centuries the great highland basin around Mexico City fed a large human population, chiefly on fish and water-beetle egg masses from the lakes and on vegetable crops raised nearby. Water for these enterprises filtered reliably from forests on the higher land. On its way down, it supplied the needs of terraced gardens on the foothills encompassing the city. Lower still, the people dug canals in the margins of the shallow lakes and gained in this fashion both a comfortable means of access to their wetland gardens and a convenient supply of water for irrigation. Sediments removed annually from the canals to keep them open served as fertilizer for the gardens. This abundance of food and water seems to have given the Aztecs enough vitality to become the overlords of Mexico.

The Spanish conquerors demanded slaves and charcoal as well as food. The hilltop forests became charcoal. Soon floods poured down upon Mexico City from the bared, water-shedding hillsides. Lake levels fluctuated wildly, one month high enough to inundate city property, the next so low the fishing boats could not reach their docks. About 1900, engineers solved the problem by draining the lakes, exposing valuable land for urban construction. Today the city is so short of water that domestic supplies must be rationed. Water flows into the pipes for only two hours a day, serving each of the twelve districts in rotation. During those two hours the householder's rooftop tanks fill with the allotted amount of the precious liquid.

Most of the water gardens around Mexico City have become streets and lots. All but a few of the hillside terraced gardens have eroded out of existence. The dry lakes are dust bowls, and the city's people depend upon food hauled to them from other parts of the country. The Spainards' demand for charcoal may have been the costliest tribute they exacted during their occupation. The levy will continue until modern knowledge is used and the cleared tree country is restored to play its former roles.

Managing the world's forests for "the greatest good of the greatest number in the long run" will never fit a single, simple pattern. Possibly man expects too much. Perhaps he is overlooking nature's generosity and being stingy in return. Dr. Sears leveled this criticism on the basis of his experiences raising poultry in Ohio over a ten-year period. The year when he gave up fighting hawks and foxes convinced him. He let the chickens range wherever they wanted to go on his wooded property, getting minerals, vitamins, and proteins for themselves instead of from the feed trough. Hawks and foxes killed some of his birds, and the population shrank noticeably. Yet at the end of the season he found the survivors to be of higher quality than ever before. Their combined weight exceeded what he had achieved in years when more chickens survived. Even running a chicken farm in natural balance with a wooded environment had proved its worth.

+O+O+O+O+O+O+O+O+O+O+O+O+O+O+O+O+O+O+O+O+O+O+O+O+O+O+O+O

CHAPTER · 17

"The Greatest Threat to Life on Earth"

❀

Insects run a disassembly line of extraordinary scope, with a cost to man that is astronomical. From all the food we grow, they take at least a tenth. As "nature's lumbermen" they destroy more timber every year than all the forest fires and fungus rots. And ever since the turn of the century, when Ronald Ross earned a Nobel Prize by discovering that mosquitoes carry malaria, the list of diseases transmitted by insects has grown steadily. It now includes many maladies of man, of other animals, and of most kinds of crop plants.

Little by little we accumulate new pests. "One World" for man affords new combinations of food, insects, and disease which make trouble for the human race. We introduce foreign insects to local vegetation and foreign vegeta-

234

tion to native insects. The European corn-borer meant nothing in a maize-lacking Europe. The striped Colorado beetle had no significance until "Irish" potatoes were planted within its reach.

Insects and fungi do so much damage to man's health, crops, and possessions that control of them is essential. Yet his remedies may be worse than the ills they are intended to cure.

Use of chemical treatments against insects began by accident in France. The wine-makers of Bordeaux were bothered by boys who entered the vineyards and stole the luscious grapes as soon as the red color showed clear and dark. To deceive the boys, the growers concocted a gelatinous mixture of copper sulphate, lime, and water, and sprayed it on foliage and fruit alike. Not only did the material conceal the ripeness of the fruit and discourage theft; it also repelled leafhoppers and diminished losses from fungus diseases. Soon the rest of the world heard of the discovery and began using "Bordeaux mixture."

At first most of man's remedies were simple. To kill the young of potato beetles, he adopted the dye called Paris green, a compound of arsenic trioxide and copper acetate. He discovered the value of lime-sulphur sprays for caterpillars, such as those of apple-destroying codling moths. Lime-sulphur was replaced after about a dozen years by a superior poison, lead arsenate, invented in 1892 to save America's shade trees from the immigrant European gypsy moth. Oil poured on ponds killed mosquito wrigglers until 1931, when Paris green was found to do a better job. Most of these materials were cheaper, although less effective, than the "botanicals"—plant extracts such as nicotine from tobacco leaves, pyrethrum from certain daisies, and rotenone from the roots of Far Eastern and South American plants related to clover.

Nicotine is one of the most dangerous poisons known,

yet it has no value as an insecticide until it is concentrated and purified. Tobacco leaves are just as susceptible to insect attack as any other foliage, and, to protect them, tobacco-growers began spraying with lead arsenate. Both lead and arsenic are highly toxic to man, but tobacco-consumers were not safeguarded from these substances in tobacco smoke because in 1906 tobacco had been declassified as a drug. It is not a food either, and therefore not covered by the Pure Food and Drug Act.

Arsenical insecticides entered the tobacco leaves upon which they were sprayed. Also, accumulating in the soil of tobacco fields, they were taken in through the roots and spread through the plant, adding still more to the arsenic content of the harvestable leaves. From 1932 to 1951 the arsenic per cigarette increased from 12.6 micrograms to 42 —more than three hundred per cent. At the same time, cigarette consumption in the United States rose threefold. Thus, nine times as much arsenic was distributed in cigarettes, vaporized, and inhaled in cigarette smoke at the end of this period as at its beginning. By contrast, cigarettes of Turkish or Macedonian tobacco still have an arsenic content of only .81 micrograms; arsenical insecticides are scarcely used in these areas.

According to Nobel Prize winner Dr. Otto Warburg, the human race has special reason to be concerned about inhaling arsenicals, whether in tobacco smoke, in the fumes from burning coal, petroleum oils, and natural gas, or in the dust from wear of synthetic-rubber tires, road tars, and similar products. Arsenic compounds interfere with the respiration of living cells and can cause cancer. In the same years that arsenicals turned up increasingly in cigarette smoke, arsenic compounds in fumes and dust grew more common, and lung cancer increased two hundred per cent in women and six hundred per cent in men, with cigarette-smokers showing a rise in lung cancer of as much as nine thousand per cent. How much of this is due to tars formed through incomplete combustion of tobacco is unknown. Most tobacco companies are striving to reduce the tar, although arsenic continues to be a hazard. One new cigarette on the market in 1959 was advertised in recognition of this fact: "NO tobacco tars, NO nicotine and, more important, NO arsenic!" The brand (Vanguard) contained no tobacco either. The poisons used on tobacco fields have turned on man.

Lead arsenate and Paris green find less use today, although they were not replaced because of their danger to the human race. They simply ceased to be effective enough against some insects. After lead arsenate had been used for less than thirty-five years, apple-growers in Colorado suddenly realized that the codling moths had got ahead of them. The caterpillars survived ten or twelve sprayings a year, whereas in 1900 two annual sprayings had given good control. These insects had become resistant to the poison, and also to an assortment of others. Some orchardists found themselves unable to raise marketable fruit, and gave up their businesses. In Washington, Virginia, and other states the same resistance to poisons was encountered. Insects were changing.

Even the most potent chemicals grew less effective. In California the citrus-growers had kept their trees free of scale insects by placing canvas tents over the trees and fumigating the insects with hydrocyanic acid—another of the most dangerous poisons known. By 1930 the scale insects grew resistant to the cyanide gas; as few as three per cent died of a standard treatment instead of the former ninety-odd per cent.

Chemists set to work synthesizing insecticides and in 1942 came out with dichloro-diphenyl-trichloro-ethane, now familiar as DDT. It was cheap and amazingly effective, seemingly tailored to the chemical nature of an insect's body. If a fly stood on a thin film of the poison, enough penetrated through its feet to be promptly fatal. The insecticide could be sprayed on walls, dusted on crops, or used to destroy mosquito wrigglers in ponds.

DDT became available just in time for use in World War II. Fleas and lice—companions of soldiers and displaced persons who have little chance to change clothing and keep sanitary—ceased to be a problem. Civilian and military personnel could be kept free of these insects merely by blowing DDT dust into openings of their clothing. Experts on military medicine sighed with relief, for they dreaded fleas and lice. Fleas could bring a recurrence of bubonic plague, the "Black Death" of the Middle Ages. Lice transmit typhus fever, "Red Death," the scourge credited with causing Napoleon's retreat from Moscow, among other pseudo-military defeats. Lice also carry trench fever and relapsing fever.

DDT could be sprayed from aircraft over whole islands in the South Pacific, wiping out all mosquitoes and ending any concern over malaria, dengue, filariasis, and other tropical diseases. Similar techniques applied to forested areas in the United States and Canada were expected to eliminate the gypsy moth and other foes of lumbermen. DDT killed the bark beetles that carry Dutch elm disease from tree to tree; surely this would save New England's American elms. Orchardists tried DDT as a coating over growing apples, anticipating that at last they would be free of codling-moth caterpillars.

Codling moths succumbed to DDT. So did enemies of the two-spotted mites, and these diminutive sap-sucking relatives of spiders became important pests for the first time. They seem immune to the poison, for their feet and bodies differ chemically from those of insects and do not pick up DDT. Mites began destroying foliage wholesale, weakening trees so much that they produced poor crops of apples.

Then DDT-resistant houseflies appeared in Italy (1947), California (1948), Illinois (1949), and elsewhere. In the laboratory, scientists studied them and found that they had really changed. They tolerated not only DDT but other chemicals as well, and mixtures of many among the previously effective insecticides. Over a seven-month period the successive generations raised in captivity increased their immunity from three-fold to two-thousand-fold. Apparently the flies would soon be able to denature any poison as fast as it entered the body. Chemical weapons were losing their value.

Today the old types of mosquitoes are being replaced by new ones immune to DDT. They are spreading from Italy, India, Florida, and California. DDT-resistant lice were encountered in Korea. Despite all our new insecticides, the codling moth, the two-spotted mite, the gypsy moth, the scale insect, and the Colorado potato beetle are still with us.

There is nothing miraculous about the way insects stepped up their ability to break down poisonous molecules and render them harmless. Human organs, notably the liver, do this all the time. Insects merely took advantage of their prolific reproduction and short life span, enormously speeding up the process of evolution by which each species adjusts to new conditions. Since 1900, when the study of inheritance really got started, man has had two generations in which to develop further. In the same period, fruit flies have gone through as many generations as mankind since 57,770 B.C.!

Every time any animal reproduces by sexual means, new combinations of parental characteristics come together. In addition, a small fraction of one per cent among the offspring show entirely new features through a hereditary trick called mutation. If mutation produces an insect that has greater ability to destroy poisonous molecules than its fellows, this one has a wonderful advantage. It, and others like it, are the ones that survive to reproduce and pass on the new ability.

Man's chemical warfare on insects provides just the kind of test for which these creatures are especially fitted. Poisons are routine hazards to animals that have adjusted and survived for three hundred million years. By contrast, no fly-swatter ever turned an insect able to escape at ten miles an hour into a jet-propelled body capable of twenty. Man needs to use more ingenuity in developing "conventional" weapons.

Real progress is possible through more insect-proof methods of agriculture and storage. Man can beat termites by building with metal or plastics or ceramic materials. Synthetic fibers have already gone a long way toward eliminating clothes moths. The incinerator and garbage-grinder leave little an insect can use.

We are prone to overlook our natural allies among insects and insect-eating birds and bats. It was only in 1959, after aerial spraying of Maine's balsam-fir forests proved unsuccessful in controlling the devastating woolly aphid, that entomologists began releasing the West German beetle *Laricobius erichsonii* in areas of high infestation. *Laricobius* was ready for the job; it used its wings and legs to reach sites on the trunks of balsam trees where the aphids (plant lice) sit, mouth parts pumping sap. DDT sprays from above had come nowhere near them.

Beetles, especially ladybeetles and their young, consume vast numbers of plant lice. Ground beetles thrive on caterpillars, including those of pests in gardens and fields. Birds reduce insect numbers spectacularly. The mosquito population drops when swallows and bats are active. Nuthatches and downy woodpeckers account for between eighty and ninety per cent of the overwintering apple worms in New Hampshire; disease and exposure to cold kill an average of just over two per cent each; parasites and man's poisons stand between the surviving few and damage to the fruit. Living agents of biological control, unlike chemical compounds, can change gradually and automatically along an evolutionary course that keeps up with harmful pests.

Other technological weapons of modern man affect both foe and friend. Every nuclear explosion increases the world's radioactivity. And it is known that radiations, whether from X rays or fall-out, increase the mutation rate. This speeds up the evolutionary process in which insects already outdo man. A nuclear attack and its all-out retaliatory action might end civilization, wiping out man as well as most of the vertebrate animals on land. Some insects —ants and the slow-growing young of the seventeen-year cicada, for instance—would have their own bomb shelters. And if the attack came in winter, the hibernating insects and the cave bats might survive, to emerge into a strange springtime after the fall-out radiations had decreased naturally, perhaps to a tolerable level. When the dust had settled, no doubt insects would be on the wing again, survivors of a still more violent poison of man's devising.

Repeated failures with chemical warfare should have shown us that no poison can ever be a panacea. Yet in the United States alone, more than six hundred different kinds of injurious insects cause annual losses exceeding four billion dollars. Agriculturalists and foresters are unwilling to accept this as inevitable, or to rely on biological control.

In many parts of the world, DDT in concentrations that can be spread economically is no longer effective for houseflies, mosquitoes, cockroaches, bedbugs, fleas, lice, or such caterpillars as those of codling moths. Other and more potent insecticides are used, usually with preference for those which are chemically stable—retaining their poisonous qualities for months or years. During 1957 a total of 65,000 tons of chlorinated hydrocarbons (such as DDT, dieldrin, and heptachlor), 22,500 tons of copper sulphate, 17,500 tons of arsenicals, and 2,000 tons of organic phosphate insecticides (such as parathion) were used to protect agricultural and forest lands, to abate the nuisance of biting insects, to control disease-carrying insects, and to safeguard stored foods. The materials, labor, and equipment cost billions of dollars. Usually the poison was judged entirely upon its effectiveness in killing specific insects in the field, without regard for other animals affected.

Chlorinated hydrocarbons, including DDT, are stable enough to be cumulative. Although spraying at the rate of two pounds to the acre may have no immediate effect upon birds, the poison builds up year after year if repeated. Birds are poisoned by three or more pounds to the acre. In one area of Patuxent National Wildlife Refuge, nesting birds decreased by twenty-six per cent after four successive annual treatments at the rate of two pounds to the acre. In another, the five commonest kinds were sixty-five per cent fewer than before the insecticide was used, with insect-eating warblers and wrens reduced by eighty per cent.

So many birds, particularly nestlings, died within a few weeks after Princeton, New Jersey, applied DDT to save its elms from Dutch elm disease that foresters recommended a new course in future: spray in early spring before the arrival of migrating birds, and then not again until the middle of July, when young birds would be large enough to tolerate the poison. This routine was followed in 1957 in and around the campus of Michigan State University. The university's robin population, long a joy to students and faculty alike because it averaged one breeding pair for each of the 185 acres in the campus, was about ready to migrate south when the DDT mists were applied to the elms at the standard concentration of six per cent. No birds were affected—at that time. But when the robins returned in the spring of 1958, they began dying on all sides. No nest on the campus proved successful, and the only robins seen there after midsummer were three adults and one young of the year, presumably visitors from elsewhere.

Three milligrams of DDT will kill an adult robin. The birds got this lethal dose by eating earthworms, always a favorite food on the campus. The earthworms had fed on fallen elm leaves, which were coated with DDT. The worms concentrated the poison in their bodies until a robin needed to eat fewer than a hundred to be doomed. Yet the elm treatment at Lansing, Michigan, had followed the current recommendations to the letter. Elsewhere, efforts to control the elm disease have left as much as 196 pounds of DDT to the acre in the top three inches of soil!

To learn how far DDT can be passed along, one group of scientists took alfalfa that had been sprayed with the insecticide, fed it to cows, milked the cows and churned the cream, fed butter from the cream to rats, then killed the rats and tested for DDT in the body fat. The poison was recovered there in substantial amounts, its toxic properties intact.

Officers of the U.S. Fish and Wildlife Service saw such destruction of fish and fiddler crabs—important foods of marsh birds—in the swamps of Delaware from aerial spraying with DDT that they tried to block all spraying of marshlands with even one pound to the acre during a three-million-acre campaign against gypsy moths in an arc from Long Island through northern New Jersey, eastern Pennsylvania, and central New York. DDT has been seen to be especially deadly to young fish in nursery shallows and ponds because they eat insects and crustaceans loaded with the poison.

In recent years protests from many quarters have followed the announcement of each new plan to apply insecticides on a wide scale. A case taken before a federal judge in Brooklyn, New York, in 1958 was decided in favor

of the defendant, upholding the right of state and federal governments to order the spraying of DDT aerially for control of insects without permission of landowners affected. Although the defense won the case, spraying was postponed or curtailed in many New England states out of respect for the sentiments of the public.

America's outstanding authority on birds, Dr. Robert Cushman Murphy of the American Museum of Natural History, has exerted his influence toward stopping the wide use of poisons on plants and soil. The National Audubon Society and many other conservation groups have lobbied vigorously against poisoning programs, regarding them as "the greatest threat to life on earth." Under this concerted action, even the terms applied to the poisons have changed. Formerly the insecticides, fungicides, and herbicides were grouped into the more comprehensive category "pesticides." Now a new term has appeared: biocides—"killers of living things." It was used in the accounts describing the hasty withdrawal from the market and frantic testing of cranberries before Thanksgiving of 1959, following the discovery that some growers had misused a weed-killer on the bogs and tainted the fruit.

Biocides were encountered again later in the same year when several shipments of chickens were condemned and destroyed because they were found to contain traces of a dangerous chemical. These actions were in keeping with the 1958 ruling of the United States Supreme Court that the word "harmless" in the Food, Drug and Cosmetic Act of 1938 means "absolutely harmless," and that foods to which poisons have been added are "not to be used at all." Curiously, the Food and Drug Act applies to fruits, vegetables, and fish, but meat comes under the jurisdiction of the U.S. Department of Agriculture, which has less strict rules.

Public alarm over biocides worries the chemical-producers. They fear that research costs will be prohibitive if it becomes necessary to spend several million dollars and a period of perhaps five years in testing each new product before it can be marketed. Fish-canners are also distressed. They realize that salmon, our foremost food fish, moves from the oceans to fresh water to reproduce and depends upon early development in streams. The flesh and bones can be contaminated by radioactive wastes dumped in the ocean, or by DDT and other biocides draining into the streams from the land.

Saving a spruce forest from budworm can mean ruining a salmon crop in another way. Fisheries scientists have demonstrated recently that improvements in the nourishment of young salmon in their fresh-water stages correspond to larger size at the time of migration to the sea. Larger size goes with greater vigor and ability to compete in the oceans, and with better harvests when the mature fish return to the rivers to reproduce. On the other hand, when biocides from aerial spraying of forest-covered watersheds drain into salmon streams, they kill the aquatic insects on which the young fish depend. This leads to undersized, malnourished smolts going to sea, fewer survivors, and a decreased catch by the fishing industry.

Conservationists have correctly pointed to the willingness of legislatures to vote millions of dollars for application of biocides, but virtually no money for research on the broad effects of these programs. One major conflict concerned the stinging ant *Solenopsis saevissima*, which became established about 1918 at the port city of Mobile, Alabama. Although it is known as the "fire ant" from the intense burning sensation it produces on human skin, the quarter-inch insect affected so few Alabamians that little attempt was made to eradicate it. Some farmers complained that fire ants built fort-like mounds a foot high and nearly three feet in diameter, each a cement-hard hazard to agricultural equipment. If a mound is disturbed, the ants swarm out and make manual labor nearby decidedly painful. Owners of land with heavy infestations sometimes found it easier to abandon the land than to get laborers to work on it. Nor are these effects of fire ants new. In 1863 the famous explorer Henry Walter Bates told in his book *The Naturalist on the Amazons* of whole villages deserted as soon as fire ants moved in.

Entomologists are still uncertain whether the dark-reddish fire ants of Mobile mutated during the 1930's, or whether a second invasion from South America struck the same city. Suddenly the original ants were supplemented by a smaller, paler type, a far more aggressive race that even attacked and destroyed colonies of the dark form. In South America the dark kind is found in northern and central Argentina and southern Uruguay. Probably it reached the United States aboard a ship from Buenos Aires or Montevideo. The light-colored race occurs more widely in South America: in the Guianas, far up the Amazon and elsewhere in Brazil, in Bolivia, Paraguay, Uruguay, and Argentina.

Whatever the source of the paler fire ants, they posed far more problems. Although they built smaller mounds, they spread rapidly out of Mobile. By 1940 they reached Mississippi to the west and Florida to the east. In 1953 entomologists found them in Louisiana. By 1957 they were in Texas, Arkansas, Georgia, and had progressed northward in Alabama almost to the Tennessee boundary. Oklahoma and both Carolinas reported fire ants in 1958, Virginia in 1959. At first it was hoped that frost would kill them each winter. But the ants merely burrowed more deeply, staying below frozen soil.

In South America, fire ants subsist chiefly on seeds, insects such as caterpillars, and the "honeydew" they gather from living plant lice. In the United States they supplement this diet occasionally by destroying seedlings of corn and other important crops, and devour helpless young birds in the nest. Quail chicks scamper around too actively to be bothered by ants. Yet these insects reach them too, by swarming into eggs as soon as the little bird inside pips a hole for escape. Fire ants attack newborn lambs, pigs, and calves, sometimes with fatal results. And they rout field laborers harvesting cotton, corn, potatoes, peanuts, and strawberries.

Alabama began a co-operative program to control fire ants in 1937, seven years after the first official report of damage by this insect. By 1950 the entomologists had found no insecticide giving better than 90-per-cent control. Treatment was still experimental, applied locally from equipment on the ground. Only Alabamians felt much concern over the fire ant. The 1952 *Yearbook of Agriculture*, devoted entirely to insects, mentions fire ants only in connection with their attacks on quail nests.

The explosive spread of light-race fire ants in the 1950's brought these insects into painful contact with far more people, and demands reached the U.S. Department of Agriculture for emergency measures. For three successive years Congress allocated $2,400,000—$7,200,000 in all—for an eradication program extending from Texas to Florida. Between thirty-one and forty-seven thousand square miles of infested and susceptible land were to be treated by aerial spraying with dieldrin and heptachlor at the rate of two pounds to the acre. These two chlorinated hydrocarbons are fifteen times more powerful than DDT, and remain in the soil at dangerous concentrations for three to ten years.

The U.S. Fish and Wildlife Service protested immediately. Sports fishermen sent lobbies to Washington and to state legislatures. Nevertheless, the Florida lawmakers approved a bill to pay $500,000 for their share in the program. The governor vetoed the bill, recalling a State Board of Health report showing that a single pound of dieldrin to the acre of marshland destroyed the entire fish population of more than thirty species. The legislature overrode the veto, despite the further information that a pound to the acre killed virtually all larger animals in treated water except a few snails and other shellfish.

Sports fishermen pointed out that their hobby alone led to expenditures close to two billion dollars annually, a quarter of this spent in the threatened area. They deplored any move that would "risk destroying priceless fish and other wildlife in at least five states." Everyone agreed that when poisons are broadcast from aircraft, some areas are missed entirely, leaving islands from which a pest can reinfest adjacent land. Other areas accidentally receive double doses, although the first treatment is calculated to apply the highest concentration to get the best kill consistent with safety. Yet all this work involves chemicals whose long-term effects on man and domestic animals, let alone on wildlife in general, have not been determined.

Conservation departments in Alabama, Georgia, Louisiana, and Texas hurriedly published descriptions of the outcome after treating the first 6,250 square miles in 1958. In less than two days after the granules of insecticide hit the ground, "alarming numbers of dead insects, many of them important in insect control or as bird food," were killed in Georgia and Texas. As promptly, naturalists found appalling numbers of songbirds and game birds lying dead in fields and streets, woods and marshes. Quail in Alabama and Texas were decimated or wiped out. Ornithologists could count only between three and eight per cent of the number of kinds of other birds seen previously.

More than a hundred head of cattle near Climax, Georgia, died suddenly after being sprayed with dieldrin from airplanes. Brood sows lost their litters. Rabbits were hard hit or exterminated, and fox cubs were found dead in their dens; the two insecticides were recovered from the flesh of these animals. Squirrels, nutrias, cotton rats, white-footed mice, raccoons, opossums, and armadillos died in large numbers.

In Louisiana the earthworm population suffered an eighty-per-cent destruction, and anglers complained that their carefully prepared worm beds were wiped out completely. Water animals proved even more susceptible: fishes, crustaceans, even snails were devastated within two or three days after the aerial poisoning of marshes, ponds, and streams. Survivors were still dying seven weeks after the spraying ended. Fire ants, however, remained in many places.

The fire-ant eradication program, which did not eradicate the fire ant, is the largest attempt so far to use a new weapon before testing it and insuring the specificity of its application. Perhaps, as an exhibit, it will help prevent any future action of the same kind, no matter how acute the local panic.

To many people it is incredible that the program was tried at all. No mention of any serious loss from fire ants in any southern state appears in the *Cooperative Economic Insect Report for 1957*, published by the U.S. Department of Agriculture. The 1957 *Yearbook of Agriculture*, devoted to the subject of soil use, lists the fire ant only among twenty-two injurious insects for which dieldrin has proved effective. Early in 1959 the federal government scaled down its recommendation on dieldrin from two pounds to the acre to one and a quarter pounds, with no explanation of how the higher figure had been set in the first place. In Alabama, where the fire-ant problem began, the state Ways and Means Committee held hearings to learn what damage the ants had actually caused; its considered judgment, passed along to the legislature in 1959, was that the fire ant posed no threat to agriculture or livestock production; Alabama declined to participate in the federal-state-farmer program during the next biennium. At the same time, proponents of ant control by biocides in Georgia admitted that the fire ant is primarily a "nuisance."

America's beloved "eagle man," Charles Broley, who banded more eagles than anyone else on earth, attributed the rapid decline of our national bird largely to sterility from feeding on fish and mice containing biocides. Laboratory confirmation can scarcely be attempted in the case of such a large bird, but something similar has been found about quail and pheasants. If breeding quail are fed DDT-tainted grain, more than eighty-seven per cent of their chicks die within the first three months—even if they receive no additional DDT. Pheasants that receive even lower amounts of DDT throughout their growing period produce few eggs, although the fertility and hatchability and chick survival seem unimpaired. Effects of this kind, and the loss of normal supplies of insects as food, are believed to explain the drastically smaller populations of bluebirds, warblers, wrens, swallows, nighthawks, and whippoorwills seen today.

Probably mass applications of biocides should be avoided except in dire emergency, or be restricted to areas where man has altered the natural community of plants and animals for raising one particular crop to the complete exclusion of all else. Poisons for injurious insects are somewhat like antibiotics for human diseases. They tempt the practitioner to use them for quick results even before a firm diagnosis has been made. As with antibiotics, the side effects from insecticides can be disastrous. Moreover, a penicillin-resistant strain of *Streptococcus* seems a parallel to a DDT-resistant race of houseflies. Both have mutated in response to poisons applied routinely.

In the field of medicine, the need for aseptic precautions has never lessened because antibiotics were available. In the field of economic biology, a return to the old methods has been found valuable too. Water management is now seen to be preferable to the use of insecticides on mosquito breeding areas. It is better to build a dike and flood a marsh, allowing fish to control mosquito wrigglers, than to cut ditches and attempt complete drainage. Permanent ponds inhabited by fish produce few mosquitoes. Instead, they yield more fish, ducks, muskrats, and a high water table in adjacent soils. By working to restore balance in the web of life, man is using nature's weapons. He may not receive the hoped-for hundred-per-cent yield on his crops, but his future can be more secure.

CHAPTER · 18

Devastated Lands

WHEN engineers and laborers began the construction
of the great Kariba Dam in 1957, the whole world watched
with more than usual interest. The reservoir back of it
would be second in size only to the twenty-five-thousand-
square-mile lake on the Volga behind Russia's Kuibyshev
Dam. The new Lake Ue would extend through big-game
country for 175 miles upstream, harnessing the mighty
Zambezi River and furnishing hydroelectric power to a
relatively undeveloped part of Africa centered in Southern
Rhodesia.

Over an area of the Zambezi basin four and a half times
as large as our Lake Mead at Hoover Dam, public-welfare
workers began interviewing the native Batonga people.
Gradually they overcame the natural resistance of each

252

tribesman to moving from his ancestral homeland to a new farm or village prepared by the government on higher ground.

The dam gates were closed in December 1958—midsummer in the Southern Hemisphere—and the Zambezi began filling the lake. As the waters advanced, the land animals from this huge area were expected to retreat. For the few that might become isolated on temporary islands, the engineers planned relief measures. Eleven men in two boats would guide to safety any kinds that could swim. Those which could not would be given ferry service. "Operation Noah" was to be more amusing than laborious.

Neither the engineers nor their wildlife consultants reckoned on the intense loyalty to home territory shown by the individual animals, or the vigor with which those on still dry areas would defend the new shorelines against invasion by the displaced creatures swimming or ferried from inundated regions. Often the men in boats would barely succeed in driving a buck up on the lake's bank before the animal would dash back into the water and strike out for its flooded homesite. If left to themselves, buck and snake alike grew exhausted from swimming in circles, and drowned rather than invade the territory of other members of their kind. Even when the rescue parties succeeded in forcing an animal far from the new lake, it usually found no peace. It was lost, and showed this in its behavior. Moreover, it was a trespasser on land already occupied. Either the newcomer or the previous occupant of the territory had to go. Go where?

Had the lake behind Kariba Dam taken a century or more to spread, following a time scale in keeping with nature's usual changes, the dilemma of the wild animals might never have been noticed. As it was, the Zambezi filled the lake so quickly that the directors of the world's zoos had no time to negotiate with the government of Southern Rhodesia, although their facilities could easily have accommodated many thousands of the refugee animals. For most of the wildlife on the flooded area, the rising waters brought complete disaster, devastating land that had supported terrestrial creatures as well as man for countless years.

Few of man's edifices fulfill a need that continues for more than a few centuries. And once he stops maintaining them in good repair, they suffer from gradual erosion, if not from vandalism. The Kariba Dam is unlikely to last forever. One day its lake will almost certainly disappear again. If plants and animals are free to repopulate the Zambezi basin at some far future date, the new community they create on the devastated bottom lands will surely differ in many ways from the one eliminated in 1959. Silt from the backwaters of the dam will become a soil quite unlike that upon which the displaced animals lived. The fauna and flora of the new community will lack some former types. Others will be more numerous and dominant.

Our real regret is that the lake at Kariba has no large, permanent island that could serve as an isolated sanctuary, a place in which a generous sample of the plants and animals of present Africa could be helped to live into the future. An island of this sort in the Western Hemisphere, encircled when the waters of Panama's Chagras River rose behind the dam at Gatun, has become famous. It is Barro Colorado, a jungle paradise in the American tropics, set aside in the middle of Gatun Lake and now under the supervision of the Smithsonian Institution in Washington, D.C.

Probably no one will ever be sure whether the virgin rain forest on the red clay hill that became Barro Colorado Island harbored any wealth of wildlife before mosquito-control teams made possible the building of the Panama Canal and the damming of the Chagras. Perhaps the howler monkeys and capuchins differ from Old World monkeys in more ways than in being able to hang by their tails. Possibly a lowland tapir on the isthmus had less attachment to a small home territory than an elephant or a giraffe along the Zambezi. Maybe America's peccaries—the collared and the white-lipped—are less fastidious than Africa's wart hogs and bush pigs, more willing to root for a living in unfamiliar soil. It is conceivable that, as Barro Colorado Island became the only large, undisturbed refuge in a drowning forest of giant trees, wildlife from the inundated territory swam ashore without guidance and seized homesites.

These differences could correspond to dissimilarities in living conditions as well as in the animals themselves. A rain forest, such as that in Panama, seems strangely vacant—as though waiting for the animals to arrive. The troops of monkeys are far apart and high above the ground. The parrots fly in pairs, the parakeets in little flocks, but the toucans call to one another across the jungle, and the grouse-like tinamous whistle to neighbors when twilight comes. We've met several families of raccoon-like coatimundis during months in the jungle. Occasionally, at night, a herd of from twenty to fifty peccaries came by. Once we were visited by a full-grown tapir and a half-sized youngster, the two together weighing perhaps half a ton. Except for these, and ants and termites, each animal appeared alone. For most of the day as well as much of the nocturnal hours, no footstep disturbs the jungle floor.

In an African reserve the elephants and antelopes and baboons and zebras are forever crossing the road. Wherever they are protected, a person sees wildebeest by the score, hartebeest by the hundreds, and a family of giraffes behind every fiftieth clump of acacia trees. If the crocodiles are left alone, the rivers and estuaries have almost as many of them floating at the surface as a New England river has pulp logs riding quietly to the mill. The solitary hippopotamus is an outcast. No wonder these animal communities observe territorial boundaries as though they were fenced in.

Keeping a full stomach and satisfying thirst must be difficult where the vegetation is harsh and thorny, the water holes far apart. These seem to be the real factors determining the outcome of the struggle for survival. Yet a person gains an impression of remarkable productivity, as most of the larger animals are sociable, traveling in conspicuous packs or herds or troops or prides. By contrast, in the lushness of a humid jungle, each creature seems to survive through perfection in camouflage. If a moth ceases to resemble a lichen on bark or a bird dropping and spreads its scaly wings to fly across a glade, it is likely to be seized by a praying mantis that materializes from a stationary pose previously suggesting only a blotched, dead leaf. Unless the mantis dispatches the moth quickly and resumes its death-like imitation, a small bird perched inconspicuously on a high branch is likely to dart down, engulf the mantis, and whisk back to its place of solitary vigil.

The devastated jungle covered by the waters of Gatun Lake is still there, submerged, nearly half a century later. Except where a channel has been cleared for ships and smaller boats, the dead tropical hardwood trees stand unrotted. All branches that originally protruded have been pruned away by termites and fungi that destroyed the wood to the water line and stopped. The limbs end cleanly a few inches or a foot below the lake surface. They form an unseen hazard for navigation and, at the same time, a record of the lowest to which the lake has fallen while serving as the high-level waterway across the isthmus, linking great staircases of ship locks opening to the Pacific and to the Caribbean.

A newer, broader, interocean canal, all at sea level in Middle America, has been discussed for years. It might be built across Nicaragua. Conceivably, the Panama Canal could be abandoned to the Panamanians and the jungle, its narrow locks to become historic relics only a little more impressive than those built by the French engineers who so valiantly fought disease to essay the first canal. Gatun Dam and the smaller one farther up the Chagras would crumble if shipping no longer paid for their repair. Gatun Lake would empty, and the Chagras River return to its natural task of cutting a channel between the mountains and the sea. Termites and fungi would finish off the exposed remains of the drowned forest, and new life would soon spread in from all the shores around. Scientists would vie with one another for opportunities to be on hand as observers while this colonization of the lake bottom began.

Suitable occasions for following the re-establishment of living communities on devastated lands have come in the past, but seldom since travel and recording equipment became so reliable. The most spectacular opportunity followed the great disaster of August 27, 1883, when two volcanoes exploded simultaneously on the small, uninhabited, tropical island of Krakatau, in the Sunda Strait between Java and Sumatra. People on Rodriguez Island, nearly three thousand miles distant, heard the sound of the concussion. So did others slightly closer, in Ceylon and Australia. No other sound waves, not even of thermonuclear explosions, have carried so far with such intensity. Krakatau's shock waves went three times around the world in every direction, affecting barometers for days. For the rest of 1883 and part of 1884 the whole earth was treated to magnificent sunsets. Extraordinary afterglows during November and December showed how much fine dust from the eruption still had not settled out of the upper atmosphere.

On Krakatau itself the volcanic cones were obliterated entirely, and a depression eight hundred feet deep in the ocean marked the site where one of them had stood. Although the nearest inhabited island was at least twenty-five miles away, the tidal waves following the explosion took at least 36,417 lives, almost all by drowning. Not only Krakatau but two adjacent islands as well were devastated by the hot debris falling on them from the sky.

Two months later, when men first risked approaching the remains of Krakatau, the island was steaming from a recent rain, showing that temperatures close to the boiling point persisted deep in the covering of loose pumice. This debris was washing away rapidly, but could be seen to vary from twenty to two hundred feet in depth.

Nine months after the eruption the French botanist E. Cotteau visited Krakatau and looked carefully for signs of plants and animals. He found one lone spider spinning a web. Otherwise the island proved lifeless. So was Sebesi, the nearest neighboring island, twelve miles distant.

A party of scientists, organized and led by the Dutch botanist Dr. Melchior Treub, came from Java in 1886 to explore Krakatau. By then, coastal plants had taken root along the shores from seeds that had arrived from remote islands with the waves. Farther inland, ferns and grasses were growing well. Reproductive parts of these plants had been carried to Krakatau by birds and by wind.

Ten years passed before another group of naturalists stepped ashore. Botanists then found the island well clothed with young beefwood trees (*Casuarina*) and fringed with half-grown coconut palms. The thin soil supported four species of orchids, the minute seeds of which cling to the feet of birds or blow in the wind. The undergrowth included a few fruit trees appreciated by man, such as wild figs and pawpaw. No doubt their seeds had been carried to Krakatau undigested and unharmed while passing through the alimentary canal of migratory birds. Insects and spiders were everywhere. A few lizards and snakes roamed the island, the descendants of some that must have reached the island on rafts of vegetation pushed to Krakatau by the prevailing winds.

The Krakatau volcano remained dormant from 1883 until 1927, when a new cone arose as a miniature island—Anak Krakatau ("Child of Krakatau")—alongside the larger one. Two years later it exploded. No lives were lost this time, and devastation was minor. But the eruption deposited an island of debris in the deep hole formed in 1883.

The World Almanac and Book of Facts for 1959 lists both Anak Krakatau and Krakatau among "the principal active volcanoes of the world." Anak K., only 520 feet high, had its latest eruption in 1950. Krakatau itself, 2,667 feet in elevation, was active in 1953. Volcanologists cannot predict how many more people will lose their lives or how much more devastation will come because of these bits of land in Sunda Strait. Biologists, however, are confident that, no matter how violent the eruption, it will merely wipe the islands clean and prepare them for fresh colonization, a new slateful of plants and animals whose lives will intermesh into a delicately balanced community.

It is easy for North Americans to be complacent about volcanic destruction. Mexico's Parícutin grew to a height of 9,100 feet in 1943, but already is listed as dormant. Boquerón, which erupted violently in 1955, is on a remote island of the Reville Gigedo group in the Pacific Ocean off the coast of Baja California. Both of these seem far away.

Northern California, much closer home, has known volcanic activity during the present century. Lassen Peak in the Cascade Range began to steam and rumble in May of 1914. A year later, glowing lava appeared, spilling through a notch in the rim of the ancient crater and creeping down the mountainside. It had not traveled far by nine thirty on the morning of May 22, the moment when the whole mountain seemed to blow up. A tremendous blast rent the top, scattering great quantities of dacite, the youngest igneous rock in the United States.

The blast of hot rocks and gas melted much of the heavy snow left from the preceding winter, releasing a deluge of hot water that mixed with volcanic dust to form a fluid mud. Down the slope the material rushed, into evergreen forests covering the lower elevations to the east and toward grassy meadows where ranchers summered cattle. Seven square miles of this, containing five million feet of timber, were swept away in a few minutes. Trees three to five feet in diameter were torn out by the roots, or snapped off. Boulders weighing twenty tons floated on the mud stream. They were deposited five and six miles away in the valleys of Hat Creek and Lost Creek, for Raker Peak caused the flow to divide into two streams. Few human lives were lost, but the havoc on the land was so great that the "Devastated Area" is still one of the chief attractions at Lassen Volcanic National Park.

Three years later another series of explosions tore at Lassen Peak. By 1921 all activity ceased; the mountain was again "asleep." It slept quietly while, on a recent trip, we picked up a few chunks of Lassen's dacite and marveled at its solid body, its superficial resemblance to some kinds of granite. It is shiny black with many white, crystalline inclusions. Other igneous rocks thrown out by the same explosions are pink and gray. Many lie where they fell more than four decades ago. Between them new trees are rising. A pond in the devastated area is almost choked with the carnivorous, insect- and crustacean-catching bladderwort. This submerged plant was in full bloom at the time of our visit. Over the aerial blossoms, flycatchers hovered, snatching the pollinating insects as they arrived.

To see the full effect of a landscape devastated by lava, we did not have to go far. We came from Lassen Park, full of the eruption story there, traveling eastbound on highway

U.S. 28. No one warned us that ahead lay the vast Columbia lava fields, which encrust 250,000 square miles of Oregon, Washington, and Idaho, with tongues licking into smaller parts of adjoining states and Canadian provinces. Of all the states in the Union, only Alaska and Texas have an area greater than this one lava field. The margin by which Texas exceeds the Columbia lava is trifling.

We emerged from the tall timber along the McKenzie River and left behind the famed fishing camps of central Oregon just as the July sun cast long shadows. Sunset colors suffused the sky with gold that brightened the rising road ahead. Eastward, we knew, lay the Cascade Range. To our right, the volcanic Three Sisters threw back the slanting rays from lacy, peaked caps of snow. Night already nibbled at the valleys.

The radiance became orange while our car surmounted one slope after another. Suddenly the land went black. Even under the blaze of the setting sun, great shoulders of dark lava reflected no visible light. On every side they thrust upward, hemming in the tortuous trail to its very ditches. The sky turned a fiery red, and the scattered pink puffs of cloud grew ominously black like smoke—a perfect match for the rocky waste through which we wound. Never had a lava field seemed to us so malevolent as under that lurid sky. We almost expected acrid vapors to swirl across the road, and to see a still molten tongue creep in dull red glory from some fissure.

Our headlights led us through this most extensive lava field in the United States, and to a little park near the city of Bend. There we completed the eerie evening by pitching our tent at the base of a giant cinder cone—511-foot Pilot Butte. Against the starry sky, the volcano loomed like a monstrous shadow from our campfire—an image of the tent silhouetted on the dome of night.

By day a lava field retains its fascination, but casts aside the nightly cloak of implacability. In many places the land remains almost as barren as when first the molten streams solidified in place. The cold surface has cracked in deep fissures, in some of which a little soil has begun to collect. In our southwestern states these badlands support cactus and quite a few kinds of herbaceous plants.

At a few places in the world the pace of colonization by plants and animals can be recognized where roads cross successive lava flows. Between Goma and Saki, on the north shore of Lake Kivu, is a spot made memorable by three eruptions in the present century from 11,384-foot Nyiragongo, one of the two nearby major volcanoes of the Belgian Congo. Our 1959 diary has this entry:

> Signs along the road warn people not to enter the lava fields, and proclaim them to be parts of the extensive Parc National Albert. Other signs give the date of each flow: 1912, 1938, 1948.
>
> The 1948 lava is so little eroded that its flow pattern can be traced easily on the surface. The only vegetation appears to be an assortment of lichens and a few ferns of small size.
>
> The 1938 flow bears grasses and some shrubs. [Appended note: collectors of orchids have disobeyed the warnings and found their favorite plants flowering in pockets where a little soil has collected. This flow was a big one. It extended across a bay of Lake Kivu, isolating a large pond behind it and cutting off access to the previous port of Saki.]
>
> The 1912 flow is harder to see, for trees and a host of bushes conceal it. When will the next eruption send fresh devastation over the countryside? The smaller volcano, 10,023-foot Nyamlagira, spewed forth a new flow in 1954, but it went in a different direction. People here, whose lives might

be disrupted or lost through further volcanic activity, talk of something else.

No *Keep Out* signs mark the lava fields in our southwestern states, and we have enjoyed exploring them, although the razor-sharp edges of the small holes in the rock surface cut our shoes to pieces. Volcanic gases once filled these cavities. Now air in them allows the lava to act as insulation, retaining the cold of winter, defying the summer sun, and reducing evaporation within the many sinkholes. Snow or rain that falls during the cooler portions of the year may accumulate as ice and remain frozen through the summer wherever the roof of a ragged pocket in the lava keeps out the sun.

At various points throughout our southwest, "ice caves" are marked on road maps. Commonly the ice is stratified, with dust marking the limits of the various years. Like tree rings, or the history written in clam shells, fish scales, and ear stones, the record of the past is preserved, layer after layer. If the sun can slip a slender pencil through a hole in the roof, it writes into the ice a little temporary pool. Green cells of simple algae thrive in the water, although they freeze each night. In the ice can then be traced the exact areas reached by the sunlight at different seasons of the year. As the sun's elevation in the sky changes, month by month, so too does the form of the green pools melted in the ice caves.

Many sinkholes in New Mexican lava fields are broadly open to the sun, and as much as an acre in extent. Most of them contain water, making possible in the midst of the lava flow the existence of a great variety of living things. After driving along a highway under a burning sun past miles of bald, eroded hills dotted with gray-green sagebrush and scrubby chaparral, the sight of a pondful of cattails seems a mirage. There, a dozen feet below the general surface of the lava, will be a pond with water-striders. Red-winged blackbirds call cheerful and familiar notes, and kill-deers scurry over the rough rocks with plaintive calls.

, The height of incongruity is to pick a rotting stem of cholla cactus from an alga-filled pool in the lava and see a shower of flat, shrimp-like scuds (*Gammarus*) kick their pale bodies out from between the woven woody strands. Or an olive-green tree frog, many miles from the nearest tree, may climb with adhesive disks on the tips of well-spread toes, approach the top of a tilted block of dark lava, and flick out a sticky tongue to capture an unwary damsel fly. The pond community has come into a state of balance too.

For a fair number of lava flows in the west, a time scale can be matched to the recolonization. Scattered through many encrusted fields are vertical openings that mark the sites where giant trees grew before the earth spewed forth the oozing, molten rock. Lava flowed around the trees and, no doubt, set them blazing. But in many places the rock mass was viscous enough to solidify where it pressed against the torch-like trunk. Some of the large trees turned to charcoal. Most of them eroded away until only the cast of the bark shows what manner of obstruction produced the present well-like holes. Yet in a few the charred remains were covered up and preserved. Painstaking scientists have removed the charcoal, measured the rings of growth still showing in the heartwood, and compared these with living trees and with beams in old Indian houses. The flow near the present straggling town of Grants, New Mexico, is an example dating from about two thousand years ago.

Erosion of broken vesicular lava into crumb-sized fragments requires a few centuries in well-watered lands near the equator, and a few millennia in our arid, temperate west. The fine volcanic matter then aids agriculture. Ethnologists are satisfied that the prolonged droughts of the thirteenth

and fourteenth centuries drove the American Indians from their cave-like dwellings at Mesa Verde and other parts of the western states, and turned them into pueblo people—still with a cave-like ceremonial room (*kiva*) to represent the former ways. When the drought ended, the Indians failed to return to their cliff houses because they had discovered that the old volcanic soils around their new homes would yield good crops. Eroded lava satisfied their needs by helping the soil hold water and keeping it open enough for air to reach healthy roots.

In Honduras we encountered this same idea while chatting with Dr. Wilson Popenoe, then director of the agricultural college at Zamorano for Latin American boys. Dr. Popenoe assured us that if the students could learn to raise good crops on Honduras's soil, they would manage magnificently when they returned to other Latin American countries. Honduras, he told us, has had no volcanic activity for so long that the soil is barren. All the other parts of Middle America have had lava flows within the last thousand years. Wherever the lava has eroded thoroughly, the soil produces wonderful crops. These were the kindest words we had ever heard spoken of volcanoes.

In recent years foresters and wildlife managers have begun to see some good in another of man's terrors: fire. When properly timed and controlled, fire can not only remove part of the litter that has accumulated on the forest floor and eliminate some competing seedlings, but also serve to open the heavy-bodied cones of lodgepole pine, pitch pine, jack pine, and some others, releasing their seeds as nothing else will do. Forests of these types depend to some extent upon occasional fires for survival. Without burning, the pines are replaced by other, shade-enduring types of trees, such as spruce and fir. Longleaf pine in our southern states seems to need recurrent fires at two- to five-year intervals, both to open the cones and to help control brown-spot disease.

Animals too may profit from fires. Recently Dr. Raymond B. Cowles, a conservation-minded naturalist, has questioned whether the taboo against burning the dry uplands of California may not be starving the rare condors, despite all the measures being tried to keep this bird from becoming extinct. According to Dr. Cowles, an occasional burn in the chaparral prevents this low woody growth from taking over the landscape and turning it into a faunal desert. After a burn he has noticed that the plants regenerate, furnishing food for a vastly increased number of animals, and also that the springs and streams increase in flow and the land becomes attractive to condors as a place to alight and feed.

New ideas on the value of fire have been put to use in marshes along the Gulf of Mexico. Employees of the U.S.

Fish and Wildlife Service deliberately burn large areas of swampland in winter, eliminating the tall dead reeds and exposing fresh growth to hungry ducks and geese. Blackbirds benefit also. They have learned that a column of dense smoke rising from the marsh marks a region offering wildly flying grasshoppers and other insects. From miles away the birds rush to the fire and snatch free meals, even flying through the smoke in pursuit of victims. Later in the spring, after the waterfowl have left, ranchers sometimes pasture their cattle on the drier parts of the burned-over swampland, letting the animals forage where each can see where it is going and avoid becoming mired.

To benefit living things, fire must be used with adequate knowledge of what is to be gained, and also constant attention. Fires started through inadequate understanding account for about nine tenths of those listed as of incendiary origin. Particularly in the southern states, where annual woods-burning has long been a tradition, forest fires are started and allowed to get out of hand. According to the statistics of the U.S. Forest Service, about a fourth of the timber-destroying blazes in the United States each year are started intentionally, but only a few of them because of malice. Careless smokers, discarding a glowing cigarette or a smoldering match as they might in the city's streets, set another fourth of the nation's serious forest fires. Careless burning of debris, careless campers who leave a cook fire unquenched, careless railroaders who let sparks from locomotives escape into dry woodlands, and careless lumbering operations together account for a further quarter of the timber blazes.

Lightning explains only 8.7 per cent of forest fires. Yet losses from this fraction are so large that in 1956 the Forest Service began a co-operative study of electric storms as "Project Skyfire." Information gained by studying the lightning potential of storm clouds and by tracing their paths with mobile radar equipment is expected to help predict where fire-fighting crews will be needed.

The Forest Service has shown remarkable progress in handling dangerous fires, reducing devastation of timberland, watersheds, wildlife, and recreational values. Steadily the areas affected have decreased: 1955, with 7,616 fires fought and 589 square miles of forest lost; 1956, with 11,730 fires and 311 square miles burned; 1957, with 6,847 fires and 167 square miles destroyed; 1958, with 10,274 fires and only 101 square miles ruined—each enumeration covering the first ten months of the year. To keep the record improving, new forest-fire research laboratories were established in 1959, one in Macon, Georgia, another at Missoula, Montana, as centers for testing improved methods of controlling sparks and flames.

After a period of dry weather, grasslands burn readily, whether on prairie or savanna. In many parts of the world, grass fires sweep across the countryside as a wall of flame under a dense pall of smoke. With a wind to help, the blaze becomes a terrifying spectacle, racing along almost as fast as a horse can gallop. Yet the effect on wildlife may be less than many people expect. Dr. Walter E. Howard, a range-management specialist in the College of Agriculture at the University of California, found that a great many creatures manage to survive even fires that burn longer and hotter, fueled by an abundance of dry bushes standing among the grass. The morning after a grass fire scorched two and a half square miles of Madera County, near Yosemite National Park, a pool of water at a spring almost in the center of the blackened tract quenched the thirst of 321 valley quail, representing all ages, in a three-hour period. During the same time thirty-one cottontail rabbits (one singed by the fire), twelve gray squirrels, and four ground squirrels took their turns. Somehow these creatures must have found refuges, perhaps in holes or between rocks, as the flames swept past them. Other birds, which might have flown away and then returned unconcernedly for a drink with the quail, included more than eighty brown towhees, twenty-seven California jays, and eighteen California woodpeckers.

How repeated burning of this kind would affect the winter forage available for deer could not be determined. Nor are pasture specialists in agreement. The issue is beclouded by differences in the degree to which grazing animals are weakening the plants, as well as variations in rainfall, summer and winter temperatures, the time of burning, and the natures of the plants.

On a year-to-year basis in western Uganda the twice-annual burning of the savanna grasses and shrubs is credited with reducing the food available to undernourished, disease-ridden Ankole ("Watussi") cattle by seventy-five per cent. Where only a single grass fire each year is the custom, a loss of fifty per cent is expected, chiefly because many of the plants already weakened by overgrazing are killed.

Approximately similar country farther south in the eastern Transvaal offers a contrast. There a late winter burn seems advisable every year in spring (October) to prevent shrubby plants from taking over the range and crowding out the edible grasses. Even in Kruger National Park and some of Natal's well-managed game reserves, experiments are in progress to learn whether the custom of the natives, who set fire to the grass as a ritual "to make it grow green again," may have some advantages if managed cautiously.

The whole community of living things must be affected by regular burning, but the outcome differs according to when the fires are started and how often. Today, in the region around the source of the Niger River in French West Africa, only a few fire-resistant savanna trees survive the annual burning of the grass. Botanists are certain, however, that forest and not grass was the original vegetation of the area. They trace the change from early times when the region was populated by pagan Negroes who cut the trees at intervals but did not cultivate the soil. The abundant rainfall let tree stumps sprout quickly, keeping grass from intruding. But later a Mohammedan culture replaced the pagan one. The trees were kept cut, the soil was cultivated until the fields were exhausted and all the stumps were dead. When the worthless land was abandoned, grass entered unopposed. If the annual burnings were discontinued now, the whole area would slowly revert to timberland. Fires today keep it devastated.

It is but a step from exhausted grasslands to desert. The famous Gobi Desert may have been forested long ago, and the sand dunes of the vast Sahara may merely measure man's past lack of care for agricultural land. Scientists today cannot be sure whether the change from useful to useless came because human activities led to an alteration in the rainfall pattern, or whether the weather shifted as part of a long-term variation.

The steps that lead to shifting sands are not always evident while watching the individual particles of rock rolled up the long slope by the wind and down the dune's other side. Little by little each sand mountain flows onward like a giant wave, engulfing ponds and trees and fields. Occasionally it overwhelms a house, crushes the building, and continues. Decades or centuries later the wreck may reappear briefly before another dune submerges it.

After the last mountain of sand has moved on, the land lies ruined. The same is true when the burden has been the thick ice of a glacier. Yet, with time, the devastation ends. Countless plants and animals, as well as multitudes of mankind, now find homes where the ice sheets stood a mile thick some fifty millennia ago. No one then could have predicted when they would melt and disappear. Scientific realization that they existed in the past came as a surprise less than a century ago. Recovery of the devastated regions has been so complete that evidences of loss are apparent mostly to plant geographers.

Whether the disaster to the soil has come through a tremendous fire, a lava flow, or submergence under water, ice, or sand, hope is justified. Eventually relief will follow, and opportunity as well. As the plants and animals recolonize the territory freshly made available, they meet new challenges that mold their lives. Adversities of the environment and the company of other pioneers stimulate the evolution of new kinds of creatures. Their adaptations accumulate, making each type of life more specialized than ever before, fitting it into a new equilibrium. This is the direction all life takes, a way in which it differs from the world of the non-living.

> The destruction of the wild pigeon and the Carolina para-
> keet has meant a loss as sad as if the Catskills or the
> Palisades were taken away.
>
> THEODORE ROOSEVELT in a letter
> to Frank M. Chapman, President
> of the Audubon Society,
> February 16, 1899

CHAPTER · 19

Tomorrow's Fossils?

❁

THE néné, official bird of our newest state, is a small goose found only in Hawaii. There it is restricted to the desolate lava country between the volcanoes Mauna Loa and Mauna Kea. A decade ago the néné seemed likely to vanish altogether. Only thirty individuals were known any-where, representing flocks that once totaled at least twenty-five thousand. Now more than fifty wild nénés live on the islands. Another 150 are protected carefully in captivity to-ward the day when once more they can be free. The con-servation drive of the past decade is paying off.

Saving the néné from extinction has been a popular cam-paign in Hawaii, enjoying the co-operation of the citizenry, various private organizations, and the local and federal gov-ernments. Consequently, the islanders were overjoyed early in 1959 to learn that a néné nest had been found in

272

Hawaii National Park—even though a mongoose destroyed the eggs in it.

Nénés formerly occupied most of the isolated patches of Hawaiian land between lava fields at elevations of from five to seven thousand feet. Excessive hunting decimated them, and introduced animals continued the destruction even after legal protection was extended. Wild dogs and pigs took a large toll, especially during the one month every year when each néné cannot fly to escape a pursuer. Until a new set of feathers grows in after the annual molt, the bird is grounded and particularly helpless. Nor can it take refuge in lakes and marshes, for none are to be found in the home territory. Moreover, these four- to five-pound geese have no webs between their toes and swim poorly. Their feet are fitted to walking across the rough, slag-like lava known as "aa-type."

On the American mainland a comparable conservation campaign caught the public eye when the once common and widespread whooping cranes dwindled to twenty-one birds. Despite a 1913 law protecting these five-foot white cranes in all states, hunters and adverse weather brought the whoopers ever closer to extinction. For years they teetered on the brink of final disaster. Recently they seem to be improving their hold on life. In 1959, with the known world population of whooping cranes at thirty-eight, an additional bird was discovered spending the winter in Missouri, far from the rest in the Aransas National Wildlife Refuge on the Texan coast.

We'll not forget the day we stalked a pair of whooping cranes, eager to capture these rare birds in our cameras without disturbing them in any way. As we crouched among the shrubbery, the vigilant cock bird stood higher than we did. The black pupil of his white eye kept constant watch in our direction. Sometimes he turned his head in such a way that we could see through his bill—in one nostril and out the other. At other times he faced us and showed two pupils at once; we were in his binocular field of view as surely as any frog or fish could be.

The red crown of the whooping crane, the movements of the two birds foraging together, their mating dances suggesting that the law of gravity can be suspended, and the characteristic flight with head and legs outstretched are all on color motion-picture film. The clarion call of the whoopers can be reproduced at will from magnetic tape. These facts have been preserved. But not the corresponding features of many members of the animal kingdom that once paused briefly on the verge of extinction and then went under.

How many of our own experiences will be beyond duplication in the next generation? Each person is but one link in the long chain of human life. We tug at the past through a heritage told us by our parents, many of the tales so vivid we feel that we too were there, although unborn. Or, for some of us, books provide the continuity.

Sometimes the two of us journey far to meet some vanishing creature, realizing with a sense of urgency that it may be now or never. Probably we feel this way because of Great-uncle Jim, who recalled immense clouds of passenger pigeons. When the old man was asked about them, his face would light up. Boyhood enthusiasms flooded back. Words tumbled forth, dramatically re-creating days when the sky was darkened by the passing flocks. The roar of their wings, as they rose from the farm wood lot, surpassed the sound of a speeding train.

The wood lot belonged to the family, and provided the goal of an annual spring pilgrimage to walk through the dappled shade of fresh leaves. We remember looking up at oaks and beeches that had known pigeons, wondering why the birds were gone from the beloved "bush." At our feet, hepaticas and trilliums, bloodroot and dogtooth violet nodded in the gentle air of May, while frogs and toads and salamanders disported in shallow ephemeral pools left by melted snow.

Now the "bush" is gone too—transformed into a housing development. Tulips and daffodils from western Europe take the place of the native wildflowers. Instead of oaks and beech, the weeping willow and *Ailanthus* of China shade the lawns. And if new generations of children there are to learn that brown toads and green frogs come from black tadpoles, this knowledge must be imparted at school—with a record of amphibian voices to show what the springtime chorus was like, right where their split-level homes stand today.

Deforestation has changed the fortunes of animal life over the whole world. The ivory-billed woodpecker, recognized by the National Audubon Society as America's rarest bird, requires undisturbed forest with plenty of dead and dying trees full of beetle grubs. Territory of this type has become very scarce in the bird's native southern states, and from 1937 to 1949 no authentic record of ivory-bills came to light despite constant searching. When a few of the birds were discovered in northwest Florida, paper companies and private parties generously donated a thirteen-hundred-acre sanctuary to perpetuate this second-largest woodpecker in the world.

Ivory-bills once were widely distributed in Cuba too, but in 1957 only a dozen birds could be found, all in one part of a single province. The land on which they live is forest owned by the Bethlehem Steel Corporation and the Freeport Sulfur Company. Both of these organizations are doing all they can to help the birds, realizing that the spectacular red-crested woodpecker is fighting its last stand.

Deforestation to make more land available for agriculture, necessary to feed a growing human population, is gradually eliminating the mountaintop cloud forests of

Central America, the last refuge of the resplendent quetzals —sacred plume-bearers of the Mayan people. Guatemala claims the colorful quetzal as its national bird, and uses its name for the unit of currency. Yet Guatemala has almost no quetzals left, and the descendants of the Mayans in other lands of Middle America no longer revere the bird. For these people today the cloud forests seem only the ominous home of a poisonous snake whose ancient name, *coatl*, is familiar as part of the name of the once-dreaded plumed-serpent god, Quetzalcoatl.

Closer to home, the increasing number of people vacationing in or retiring to Florida has elevated the price of land on the keys so rapidly that the U.S. Fish and Wildlife Service is having difficulty providing refuge areas for America's smallest deer. The key deer, a race of the Virginia white-tailed, stands only about twenty-seven inches high and reaches a weight of but thirty pounds—a tenth as much as the more widespread, familiar animal.

The hurricane of 1937 decimated the key deer, and later storms limited them almost completely to Long Pine Key. There a careful census in 1955 revealed just ninety-four of the collie-sized animals, most of them in danger from illegal hunting, vicious dogs, and gradual elimination of habitat. Congress authorized the U.S. Fish and Wildlife Service to spend $35,000 to acquire up to a thousand acres of refuge land, but only eighteen and a half acres could be purchased. The fund was barely a third enough at current prices, even if owners had been willing to sell. As a stopgap, some six thousand acres were leased, and hope rests on law-enforcement and habitat-improvement to bring the small herd back to a more secure size.

For native animals, islands provide particularly dangerous homes. As the eminent conservationist Dr. Francis Harper pointed out, more than thirty-eight per cent of the kinds of mammals that have become extinct during the past twenty centuries were West Indian. The birds there have fared no better. During the past three hundred years the Caribbean islands lost eleven species and ten subspecies found nowhere else, while the world's six great continents were deprived of eight—including the heath hen (1870), Labrador duck (1878), Carolina paroquet (1904), passenger pigeon (1907, in the wild condition), and Eskimo curlew (1932). As recently as 1953 the West Indies were in danger of losing an additional twenty kinds of mammals, fifteen of birds, forty-nine of land reptiles, and uncounted kinds of amphibians and local fish. By contrast, no animal life of the Caribbean mainland (Central America) has vanished, so far as is known, for better than two thousand years. It still finds space there to get away from man.

Islands are the makers and destroyers of species. The water barrier around each one prevents the creatures living on the ground from finding much variety in mates. Inbreeding is inevitable. At the same time, the choice of habitats is limited and the web of life comparatively simple— usually with only a few kinds of predatory animals. Creatures that survive in these circumstances become highly adapted and seem less able to tolerate changes. With the arrival of man, his domestic animals, mice and rats, or introduced predators such as foxes and mongooses, the island animals tend to die away. Soon many of them are extinct.

This limitation through heritage and habitat is evident on large islands and small. Destruction of the native animal life is most obvious on Jamaica, Puerto Rico, Guadeloupe, and Martinique, where the human populations are large. Yet uninhabited tiny islets owned by the United States government in the Virgin Islands are also home to dwindling species.

The U.S. Migratory Bird Act does not apply to the Virgin Islands. Conservation is not understood there, and any law-enforcement to protect surviving animals would be so unpopular as to be almost impossible. The white-crowned pigeons, for example, are vanishing. In 1957 only three hundred of these handsome birds could be found on their sole known breeding grounds. Yet the colony is still raided annually during the breeding season. Gunners shoot the adults; eggers rob the nests. Squabs are loaded into skiffs for table use on the larger Virgin Islands. Soon this attractive visitor will be unknown in the many tropical islands from the Florida keys to the Central American mainland and down the island chain as far as Dominica.

It is easy to criticize the islanders for their shortsightedness in destroying a food resource. It is easy to quote Albert Schweitzer or Joseph Wood Krutch to the effect that men who destroy works of art are called vandals, whereas those who destroy living animals are sportsmen. But few of the islanders take life merely for target practice. The rest are intent on supplementing a diet impoverished by island living. F. R. Fosberg expressed it well: "Hungry people are not easily convinced that anything they can use should be preserved for the enjoyment of posterity."

Men charged with the safety of human lives and expensive equipment have needs almost as urgent as those of hungry people. On Midway Atoll, more than a thousand miles west of Honolulu, a major airport is operated by the U.S. Navy in the midst of a battle with 252,000 albatrosses. Neither the installation of great runways nor the almost constant arrival or departure of large aircraft has disturbed the nesting birds. Nor does an albatross, soaring with its wings stretched out seven feet from tip to tip, dodge an oncoming plane. Each thousand daytime arrivals or departures is marked by ten in which a propeller-driven aircraft strikes down a soaring bird. Collisions between albatrosses and planes have caused about $300,000 damage in each recent year, mostly to propellers and radomes. The possibility that a bird will cause a crash, with loss of lives and plane,

keeps the Navy seeking a way to eliminate the competition for airspace.

The first impulse was to purge the island of birds. This would require at least five years, for as soon as each young albatross can fly well, it soars away, not to return until it is mature. Counting those which have come back to breed for the first time and older, more experienced birds, the albatrosses of Midway Atoll include more than a third of the world's Laysan species and a sixth of the black-footed kind as well.

The suggestion that Midway's birds be slaughtered raised a tremendous storm. Too many conservationists remembered the disappearance of Steller's albatross in 1933, when the last five birds were killed on Tori Shima in June —just after the Japanese government had made the nesting site a sanctuary. They feared that the Laysan albatross could follow Steller's into oblivion if the Midway population and a few others were destroyed.

Attempts to move the albatrosses from Midway failed completely. Birds marked with a conspicuous dye were flown to Guam, to northern Japan, and to Puget Sound, four thousand miles away. Soon they were back home. Nothing frightened the parents away from their eggs and chicks. If the nests were robbed, they simply laid more eggs. Nor would more than about seven hundred birds breed on the non-strategic islands of Kure Atoll, about seventy-five miles farther west.

Navy officials decided to improve Green Island in the Kure group, laying it out during the summer of 1959 as an ideal nesting ground for albatrosses. Bulldozers were put ashore, and sixteen swaths, each fifty feet wide, were cut through the brush as leveled landing areas where albatrosses could alight easily and take off again. While the equipment was there, it was used also to build a large lagoon in which fresh water would collect, perhaps attracting still other waterfowl. Simultaneously, on Midway, additional bull-dozing cut away hills suspected of helping produce updrafts upon which albatrosses soared. Paving crews blacktopped great areas all around the airport facilities to eliminate nesting sites. Yet by October—the peak of the breeding season—these birds the Navy men dubbed "gooneys" showed no signs of transferring to a less contested spot in the mid-Pacific.

Island creatures, though so obviously vulnerable, have sometimes demonstrated that they have unsuspected nest eggs on still more remote bits of land. Species officially declared extinct have suddenly reappeared. Steller's albatross did this in 1954, when a few returned to Tori Shima to breed. The very trusting petrel of Bermuda, named the ca-how from its call, vanished by 1621 after the earliest visitors to the island took advantage of its trustfulness. Until 1951 the cahow was acknowledged to be no more. Then a few were found nesting in an outlying group of rocks which military aircraft during World War II had used for strafing practice and where bits of coral are pocked by bullets.

The cahows have been under almost continuous observation ever since.

In New Zealand, two years after a large flightless bird had been described as *Notornis hochstetteri* on the basis of a single fossil skull, two living specimens were captured by the Honorable Walter Mantell, a cabinet minister. To them the Maori name *"takahe"* was given. When they died in a zoo cage, the bird was again believed to be extinct. In 1948, half a century later, an ornithologist from the Dominion Museum recognized and succeeded in capturing two more *takahes* in the wilds of southwest New Zealand. Early the following year a whole colony of these birds, including two breeding pairs, was found and photographed. They are still on the list of vanishing species. Any creature whose numbers are so few needs only a minor natural catastrophe to bring the end.

Sometimes a misfortune for man proves to be the one change a dwindling species needs for recovery. The involvement of Japan with China and World War II had this effect upon sea otters. While human attention was elsewhere, these remarkable mammals of the North Pacific and adjacent arctic waters prospered amazingly. Biologists have concluded that sperm whales and mankind are their only enemies. Without disturbance from the latter, the sea otters came back from the edge of extinction.

The whole history of sea otters is one of persecution, especially from the discovery of Alaska in 1741 by the Danish explorer Vitus Bering until 1867, when the United States bought the area from Russia. During this period some 260,-790 sea-otter skins were received in Siberian ports, and an unknown number went to China or direct to Europe to be made into the most expensive of fur garments. In 1804 alone, fifteen thousand skins went from Sitka in a shipment valued at a million dollars. So desperate were the men exploiting this resource that in the early days they enslaved hundreds of Alaskan Eskimo men and, from time to time, murdered whole settlements in the Aleutian islands to eliminate potential slave labor for competitors.

By 1885 the sea otters had become so scarce that only 4,152 pelts were taken that year in Alaskan waters. The price per skin rose to an average of $360, making the total harvest still worth one and a half million. Soon a drive was on to get the few survivors, and the animal appeared doomed. To save it from extinction, the United States government extended complete legal protection. But poaching continued into the 1930's, with skins reaching the American market as recently as 1940 from Japan, selling for from $600 to $900 each. No real recovery for the animals seemed possible.

The U.S. Fish and Wildlife Service undertook to study sea-otter habits, searching for ways to encourage the increase and spread of the survivors. Apparently the animal needs a particular type of shoreline, one shallow enough so that kelp beds will grow well and shellfish thrive on the bottom. About two thirds of a sea otter's diet is sea urchins. Mollusks account for about a quarter, and various crabs and fish the remainder. On this food the diving mammals reach a length of fifty-five inches and a weight of fifty pounds.

The kelp seems to help them escape from killer whales, and serves also as an anchor line to which they can cling while sleeping. During the winter, sea otters haul themselves out on rocky coasts except while feeding. In summer they remain in the water continuously, playing or diving among the kelp, or floating belly up. Often they crack sea urchins and clams on their chests as though upon a table. Groups ("pods") include both sexes and all ages. They show no tendency to migrate, and for this reason spread very slowly whenever populations do increase.

Just before and during World War II sea-otter populations increased wonderfully, both in the Aleutian archipelago and down the Pacific coast as far as southern California. From an estimated total of two thousand individuals in 1937, the herds rose to a tally of about 3,420 by 1943. Probably more than double this number are alive today. In 1949 a census plane recorded 1,321 sea otters on a single transect between Tanaga and Kiska in the Rat Island portion of the Aleutians—a mere 160 miles.

The main population near Amchitka may already have grown to correspond to the amount of shellfish available as food. Further increases may depend upon whether the sea otters spread to other suitable feeding grounds. The U.S. Fish and Wildlife Service is testing this by transplanting little pods of a dozen animals or less, shifting them from the "sea-otter belt" in the Rat Islands to other points along the Pacific coast. Eventually they hope to have a few sea otters to harvest, perhaps building up a sustained fur industry comparable to that based on Alaskan fur seals.

Scientists have often suspected that large size predisposes animals to extinction. They point to the Irish elk *Megaceros hibernicus* with its 11½-foot span of antlers, to the gigantic rhinoceros-like baluchitheres, the Columbian mammoth, and the great dinosaurs. Today the situation supports the same idea. Most of the world's conspicuous animals are in danger.

Biggest of the threatened land animals are the rhinoceroses, especially the one-horned species on Java. Not only do sportsmen relish stalking the unpredictable, two- to three-ton animals, but the Chinese offer as much as $2,500 for a single horn, either to be made into a cup supposed to render harmless any poisoned drink, or to be powdered as a reputed aphrodisiac. The dried blood sells for a dollar a pound, to be applied to wounds and charm away infections. Rhino skin is esteemed both as a delicacy and as a medicine. About forty of the surviving Javanese rhinos live on a small government reserve in the western part of the island, sup-

283

posedly under protection by wardens employed by the Republic of Indonesia.

America's largest and most dangerous carnivore, the grizzly bear *Ursus horribilis*, is also vanishing. Originally the eighty-odd races of this half-ton animal ranged throughout most of the western half of the United States, the western third of Canada, and large areas of Alaska. In 1939 the U.S. Fish and Wildlife Service could find only 1,100 of these bears in the United States. A decade later the total had fallen to 755. Of these, 570 roamed wilderness areas in Montana, 120 in Wyoming, 50 in Idaho, and 15 in Washington state. Alberta and British Columbia still have grizzlies, but they have practically vanished from the plains and now are found only in deciduous and coniferous bushland and along the mountains and foothills. In 1949 Alaska was reported to have about 5,700 grizzlies.

Generally, these big yellowish-brown bears mind their own business—getting enough to eat by catching the young and sick of deer, digging out gophers and ground squirrels, spacing these meals with carrion, unwary birds, snakes, fish, insects, and a wide assortment of roots and fruits. Occasional individuals develop a liking for colts or cattle, and thereby invite destruction. This retaliation, plus the facts that their meat is to man's liking and that their pelts are worth a hundred dollars because of scarcity and size, has led many a hunter to take the annual bag limit of two grizzlies.

In undisturbed territory, grizzlies range over as much as twenty square miles apiece and are fairly immune to hunters. Given opportunity, however, they congregate at garbage dumps outside human communities, just as they patrol streams during salmon runs. While concentrated in this way, the grizzlies can be destroyed easily. To prevent a recurrence of this destruction in the part of Alberta where oil was found in 1957, the provincial government set aside seven thousand square miles as a sanctuary for the two to four hundred bears inhabiting it. At about the same time, when stock-raisers in Montana asked for government help in elimination of a cow-taking grizzly, the U.S. Forest Service saved the bear by declining to renew the grazing permit. The area of public land which is home to that particular grizzly will remain closed to stock for the lifetime of the bear.

On Kodiak Island, Alaska, as well as along the mainland peninsula as far as Cole Bay and on one or two islands of the Aleutians, lives the giant Kodiak bear, *Ursus arctos middendorffi*, much coveted as a trophy of the hunt. Many of them are protected in the Kodiak National Wildlife Refuge, and stay there except when the salmon are running in nearby rivers. Then they compete with fishermen, sometimes destroying as much as thirty per cent of the fish that have not spawned yet. This amounts to a pack value of $118,000 annually, and represents also a significant loss to propagation. On Kodiak and Afognak islands, the natives welcome any opportunity to add bear to their diet, although most people and even dogs find the flavor of the meat too strong.

The laws administered by the U.S. Fish and Wildlife Service permit about two hundred bears a year to be killed on Kodiak Island. This does not satisfy the natives, the trophy-hunters, the salmon fishermen, or those interested in the survival of the dwindling bears. Although they are technically carnivores, the Kodiak bears are plant-eaters too, grazing on a wide variety of foliage during spring and summer, eating quantities of fruit in season, and dieting extensively on seaweeds along the coasts in winter. They obviously prefer carrion to vegetation, but settle for a plant diet rather than bestir themselves enough to catch fresh meat. An exception to this occurs during the annual salmon run, when they expertly flash their claw-armed paws into the rivers and scoop out the big fish.

Bear management has progressed now to the point

where a decision can be made as to the size of bear to be "raised" on refuge land. By reducing the population deliberately, a lesser number of giant bears can be encouraged for non-resident hunters and to keep damage to salmon at a minimum. Alternatively, a greater number of smaller bears can be produced, yielding more meat for natives while adversely affecting the fisheries industry. The latter course, with more bears, might help ensure that these spectacular survivors among the world's living carnivores would not become extinct in the near future.

Among the earth's waterfowl, the giant is the trumpeter swan, a magnificent bird with a seven-foot wingspan. Formerly it migrated over many western states and Canada, flying in V formation like gigantic geese. Hunters brought its numbers down until the species appeared doomed. In 1935 only seventy-three of these swans could be found in the United States, all of them concentrated in Montana. To protect them, about fifty square miles of marshy lakes were set aside as the Red Rock Lakes National Wildlife Refuge. There the survivors apparently became non-migratory, although they do wander beyond the refuge property.

By 1941 the number of American trumpeters rose to 211, and by 1954 to 642. Conservationists sighed with relief. Complete protection was saving the swans. Then, in 1955, the total dipped to 590, of which ninety-five were young of the year. Where had the others gone? With so much publicity issued by sportsmen's groups, state and federal government offices, and conservation organizations of all kinds, it seemed unlikely that fifty-two trumpeters could have been shot. The managers of the Red Rocks Refuge took the trouble to catch one hundred adult swans, capturing them alive and unhurt during the summer molt of 1956. When examined by fluoroscope, thirteen of the hundred were found to be carrying lead pellets from shotgun shells—unmistakable evidence that hunters had wounded them without bringing them down.

Fortunately for trumpeter swans, both Canada and Alaska harbor additional populations, and these birds are somewhat less subject to hunting because of the remoteness of their migratory routes and destinations. In 1948 a bird census of Alaska revealed about three hundred trumpeters there. Canada harbors about double this number. A few pairs have been sent to England, to be studied and bred in captivity at the Severn Wildfowl Trust, where other vanishing species (including the Hawaiian néné) have received special care.

Animals already extinct are still remembered from time to time on monuments. The passenger pigeon was commemorated in 1947 by the Wisconsin Society for Ornithology, thirty-three years after the last one died in a zoo, forty years later than the final sight record of a pigeon of this remarkable kind in the wild.

From Egyptian monuments it is evident that the giraffe and the African elephant continued to live in southern Egypt until about 3000 B.C., although both were rare after 4000 B.C. African rhinos disappeared from Egypt about the same time. Much later sculptures, carved during the New Kingdom (1580–1090 B.C.), show the fallow deer and the famous auroch, a beast restricted to northern Europe by the time of Julius Caesar, and the presumed ancestor of modern cattle. Herodotus tells of otters playing along the banks of the Nile in his day—the fifth century B.C. Later naturalists questioned whether he confused otters with mongooses, but the mosaics and bas-reliefs bear out his written claim.

The rate at which animals disappear is almost directly proportional to the rate at which the human population grows in an area. The distinction between the relatively unpopulated Upper Egypt, near the Sudan, and Lower

Egypt with its multitudes of people is shown by the fact that the last known hippopotamus in Lower Egypt was killed in 1600, the last in Upper Egypt in 1868. As recently as 1830 the ostrich vanished from the Egyptian scene, and crocodiles disappeared from the Nile north of the Sudan about 1850. The wild boar, which formerly overran the entire delta of the Nile as well as the Wadi Natrun and Lower Egypt, became extinct there in the late nineteenth century. The last cheetahs survived somehow to the west of Alexandria, but were exterminated during World War II.

At present the Egyptian Publicity Office advertises migratory waterfowl from the northern lakes and marshes of Europe and Asia as essentially the country's only game— available during the winter season but not through the rest of the year. Gazelles have become rare, and so are the formerly abundant Barbary sheep, the Nubian ibex, and the leopards that formerly preyed upon them. Even the hyena, once so common, is rare and localized. The native fauna of the lower Nile basin is almost gone, and with it most appreciation there for wildlife.

The distinguished paleontologist Henry Fairfield Osborn complained of the same disregard all over the world:

> We no longer destroy great works of art. They are treasured and regarded as of priceless value; but we have yet to attain the state of civilization where the destruction of a glorious work of nature, whether it be a cliff, a forest or a species of animal or plant is regarded with equal abhorrence.

At least 107 species of mammals have been eliminated altogether since the time of Christ, and more than a hundred different kinds of birds since the demise of the last dodo in 1689. No one knows the corresponding information for reptiles, amphibians, fishes, and lesser creatures, but all the evidence points to a similar loss. Of these creatures, only

288

about a third disappeared in the period from A.D. 1 to 1801. Another third became extinct during the past century, and the remaining third in the last fifty years. The stupendous increase in the rate at which mankind is eliminating other species can be seen in information about the mammals: one kind exterminated for each fifty-five years from A.D. 1 to 1801; one lost for each year and a half between 1801 and 1901; and one species gone forever each year in the current century.

In our most civilized countries today, the statute books make a distinction between mankind and animals, meaning non-human animals. We have societies for the prevention of cruelty to animals, and separate laws for the protection of men, women, and children. We cleave to *Homo sapiens*, and regard all other species in the animal kingdom as of distinctly lesser worth. Yet we have treated man no differently. Between 1613 and 1828 the Beothuck Indians on Newfoundland were hunted to extinction. Natives on Tasmania followed the same route between 1804 and 1876. Today's surviving Bushmen in Africa and aborigines in Australia are dwindling rapidly. To save them, at least as much care is needed as is accorded the sea otter, the American bison, or the whooping crane.

Modern man tends to lavish his attention, wealth, and dreams upon machines. Residing remote from the soil, it is easy to forget that "apart from the members of our own species, [living things] are our only companions in an infinite and unsympathetic waste of electrons, planets, nebulae and suns." That we are not alone should be a consolation and a joy. The steady progress that is evolution may, in fact, depend on man's continued success no more than it did upon the dinosaurs. They too were dominant—for a while.

> A LL *conservation of wildness is self-defeating, for to cher-*
> *ish we must see and fondle, and when enough have*
> *seen and fondled, there is no wilderness left to cherish.*
> ALDO LEOPOLD: *A Sand County Almanac* (1949)

CHAPTER · 20

Sanctuary

❋

FEW benefits conferred by man upon other forms of life can compare with granting immunity from human interference. The broadest benevolence of this kind ever devised seems to have arisen in India about the fifth century B.C. as the Brahman priest caste pressed the Buddhist doctrine of reincarnation to its logical extreme. No devout Buddhist will kill an animal, not even an insect, for fear of endangering the reborn soul of a human ancestor. Nor has this superstition been supplanted, although Indians of other faiths eat meat when they can get it. Cattle and cobras alike remain protected widely. As recently as 1958 the Indian government yielded to religious pressure by banning the export of rhesus monkeys for medical research in Occidental countries. No shortage of monkeys was cited; reincarnated souls were at stake.

Elsewhere the protection of animals from human depredations has usually arisen for the exclusive benefit of a king or ruling class. On this basis the world's earliest known wildlife-management program was established about 250 B.C. by the Emperor Asoka in Magadha, northern India. He set aside a system of game refuges throughout his kingdom, arranged for maintenance of habitat suitable for a list of protected game animals, and even enforced bag limits upon the noblemen permitted to hunt with him.

Later rulers in India continued an interest in wildlife as a resource, right down to British days. Their regard for game, coupled with general Buddhist concern for life of any kind, seems to have turned the Indian subcontinent into a vast sanctuary. For two thousand years—until 1951—India was unique among the world's large land areas in having exterminated no kind of wild mammal. By contrast, North America has eliminated twenty-seven different kinds in the past 115 years.

India remains world famous for her fauna, particularly her tigers and Asiatic lions. The latter continued to thrive in the northwestern states while first the Cape lion of Africa became extinct, then the European lion vanished, and next the Barbary Coast representatives (source of supply to Roman amphitheaters during early Christian days) followed into oblivion. The Asiatic lion is somewhat smaller and less spectacularly maned than the lions of central Africa, but appears embellished with wings in terrifying images chiseled and sculptured as decorations throughout the Assyrian and Persian empires of Darius and Xerxes.

Today the Asiatic lion is making its last stand in the Gir Forest of arid northwestern India. Formerly the Gir extended over twelve hundred square miles and consisted of a dense stand of teak and thorn trees. A few people lived there, but were prohibited by religious convictions from killing any animal life, and hence did not molest the Gir lions or their prey. Growth of the human population led to encroachment and induced the British and local rulers to set boundaries to the remaining 480 square miles as a legally protected reserve. Since 1900, however, the semi-nomadic Chavan people have moved in with between thirty and eighty thousand head of half-starved zebu cattle and water buffaloes. Even without hunting, the lions appear doomed, for the Gir Forest cannot reproduce itself. Seedlings and all young trees are eaten by the domestic livestock. Desert scrub is taking over. Whether the Indian government, which has been unable even to collect the grazing tax levied on herders in the Gir sanctuary, will be able to move the remaining lions—numbering less than a hundred in 1958—to a larger, less competitive refuge and police it properly in central India remains to be seen.

The earliest known conservation effort in the Western Hemisphere affected an equally arid but far less accessible area, mostly between twelve and sixteen thousand feet elevation in the mountains of Peru. There, in the twelfth century A.D., the Incas sequestered the vicuña as a royal animal. Anyone found molesting these little golden-fleeced camels, except during one of the state-run hunts, was subject to death. This rule was but part of a code of laws protecting vicuñas and all other game in the Inca Empire.

In a sense, the democratic people of the United States and Canada have sequestered all of the game mammals and birds in a corresponding way: all of them belong to the citizens, to be used according to a code. Yet the first attempt to improve the fortunes of wildlife in North America came so inconspicuously that none of the naturalists' magazines of the day took any notice of the event. Only later generations came to appreciate that in 1852, Samuel Merritt of Oakland, California, bought a slough and developed it into a sanctuary lake attracting thousands of wintering waterfowl. Lake Merritt still serves this role, now almost at the center of Oakland's business district, showing clearly how wild creatures will take advantage of protection, food, and a resting place despite the surroundings. The lake acquired legal status as the country's first official wildlife refuge in 1870, barely two years prior to the establishment of Yellowstone as the first national park in the world.

Killing of wildlife in our national parks was permitted until 1894. By then another outstanding wildlife refuge had been established in America as "Bird City" on Avery Island, near New Iberia, Louisiana. Mr. E. A. McIlhenny, head of the Tabasco "liquid pepper" business there, decided to imitate a seventeenth-century rajah in India by building a sanctuary around his home and inducing showy birds to inhabit it. Like the rajah, he turned his property into a park replete with trees and pools and suitable as a habitat. Then he had erected a great cage of poultry netting, fifty feet square, high among the buttonwood trees, and stocked it with eight young snowy egrets. Until autumn, when they reached maturity, he fed them regularly, counting on them to regard the property as their home. When it became time for the egrets to migrate, the cage wire was removed, and soon afterward the birds left. Six of the eight returned to the feeding stations the following spring. Two pairs nested, starting a colony that has grown greatly and attracted other breeding birds of many kinds.

Mr. McIlhenny's grandson still maintains "Bird City." Even at the beginning of winter, when the bald cypresses have shed their needles, the sanctuary he showed us on a recent visit is a fairyland for all manner of wildlife. Countless reflecting pools cascade gently from level to level, each the home of ducks and wading birds. Shrubs that flower in spring and summer give a sense of privacy. The waterfowl have grown used to the men who quietly attend to the necessary work of keeping the property in order.

An entirely different paradise for birds was founded in 1909 in the American tropics a mile from the northeast shore of Tobago, in the crown colony of Trinidad and Tobago, close to the Venezuelan coast of South America. Sir

William Ingraham, who owned the *London Illustrated News*, transformed his 450-acre islet into a West Indian home for twenty-six pairs of the showy bird-of-paradise from Dutch New Guinea, just north of Australia. Nowhere else outside New Guinea do birds-of-paradise live in the wild state.

Upon the death of Sir William, his sons presented the island and its birds to the people of Tobago and Trinidad, and the wild aviary has been maintained ever since by the Forestry Department. Through the courtesy of the conscientious officials there, we were able to ride a small boat across the treacherous currents in the intervening water and spend a while in the sanctuary's paths. All through the day the descendants of the introduced birds could be heard from the hilltop, calling to one another in powerful, isolated whistles repeated two or three times in the same low key. In early morning, somewhat after noon, and again toward dusk they seemed easier to see as they approached certain locations on the island and could be recognized high in the trees. A cock bird, as large as a small rooster, would fly overhead and alight on a commanding site. Several hen birds might follow, exciting the male to display the gloriously golden plumes that arise under his wings. In courtship he raises his wings repeatedly, causing these long thin feathers to cascade upward and outward like a shower from a fountain. Such is the magic Sir William transplanted to the Western Hemisphere.

On the bird-of-paradise island, any native creature may be sacrificed to protect the imported birds. This policy contrasts sharply with that followed for Barro Colorado Island in Panama, where no foreign creature is brought in and nature maintains her own precarious balances. A person goes to Sir William's island only to see the showy birds-of-paradise, whereas many outstanding biologists count themselves "alumni of B.C.I." because they gained there a first-hand acquaintance with life in a dense tropical jungle scarcely modified by man.

Decisions on policy regarding a sanctuary of any kind are usually difficult. For which inhabitants is the reserve intended to provide special safety? The temptation is to favor the gentlest and eliminate as undesirable all "blood-thirsty" predators. The dramatic explosion and collapse of the Kaibab deer herd served to change this attitude in our national parks and wildlife refuges. But by then, most public lands had been cleared of cougars. Only Mount McKinley National Park in far-off Alaska still harbored many wolves. Coyotes and other smaller predators had been reduced almost to ineffectiveness.

In the mid-thirties the Park Service reversed its stand, and freshly held:

> That every species shall be left to carry on its struggle for existence unaided, as being to its greatest ultimate good, unless there is real cause to believe that it will perish if unassisted . . . that no native predator shall be destroyed on account of its normal utilization of any other park animal, excepting if that animal is in immediate danger of extermination . . . and that the rare predators shall be considered special charges of the national parks in proportion that they are persecuted everywhere else.

This policy has proved completely workable.

Where human predation and domestic animals are excluded and all native creatures are welcome to work out their separate destinies, a sanctuary becomes a sort of living museum. Wild predators may even be reintroduced, to take the place of those eliminated in earlier times. Mink are being freed in Tinicum Marsh, a wildlife refuge in the industrialized southwest corner of Philadelphia. Already most visitors there have forgotten the former boundary between sixty acres held by the federal government and the 145 acres contributed in 1958 by the Gulf Oil Company. One observation platform has been built at the edge, and others are planned as vantage points from which the public can enjoy watching the 230 different kinds of birds taking advantage of the natural, wild facilities. Muskrats and turtles are thriving, and over 150 types of native plants have been recognized on the marsh.

Domestic animals that have gone wild constitute a problem in many sanctuaries. The officers of the National Park Service are constantly troubled at Grand Canyon, Canyon de Chelly, Death Valley, and others of the national parks and national monument areas by feral burros, the offspring of animals left behind in early prospecting days. They destroy vegetation, befoul waterholes, and disturb the balance of nature the Park Service attempts to maintain. Elsewhere, dogs hunting by themselves often upset plans for deer and elk. House cats tend to take the place of native predators, particularly in smaller refuges near centers of population. Even cats receiving six ounces of canned pet food daily prefer to catch a food supplement of voles, chipmunks, small birds, shrews, and insects. They spend many hours hunting for this wild fare whenever let out of the house.

The populations of conspicuous animals on sanctuary grounds always tempt poachers. Most poaching is recognizable as armed robbery, often taking animals that have been raised at considerable expense according to a program of conservation or planned harvest. The poacher sets himself beyond the law, and is often willing to use his gun to resist arrest. The distinction between acts of this type and those of bandits staging an armed holdup at a bank may lie chiefly in the fact that a bank's stock-in-trade is insured and replaceable, whereas the living creatures are not. Yet some people regard a poacher as a reincarnation of Robin Hood, who lived dangerously on game animals from private estates of the rich and somehow gained respectability by dividing his plunder with the poor.

The severity of poaching depends largely on how much backing the public is willing to give game wardens appointed to police a sanctuary. The question has to be answered repeatedly, as in January 1959, when a rifle-carrying deer-trapper was surprised by a warden a few miles south of Belvidere, New Jersey, and was shot dead in an exchange of gunfire as the warden tried to take him in under arrest. The warden was booked for murder and ordered to the county jail, unable to be freed on bail in that state. Had the tables been reversed, how many people would have risen to the defense of the officer discharging his duty? We remember on the Florida coast the stone marker showing where Guy M. Bradley, the Audubon Society warden of Monroe County, was shot dead by a plume-hunter resisting arrest for poaching egrets in a Florida sanctuary. In that instance, the suspect was not brought to trial because of insufficient evidence. A mob of residents burned down the house of the suspected man, but the Audubon Society had trouble finding another warden. A second officer, Columbus G. McLeod, suffered the same fate before the feather fight reached its bitter end.

Many people seem to regard sanctuary land as an indefensible luxury, merely waste acres that need not be given a second thought if some other use for them can be found. New York state officials have shown this attitude twice toward the only national wildlife refuge within their jurisdiction. Montezuma Refuge, at the north end of Cayuga Lake, has long been famous as a waterfowl marsh, a breeding place for some kinds and a twice-annual stopping point for thousands of ducks in their southward and northward flights. Yet when the barge canal was built to allow inexpensive movement of bulky cargoes between the river systems, no

one bothered to provide the low dike needed to hold water in the old marsh. Instead, it was drained and largely destroyed. Later the U.S. Fish and Wildlife Service took over 5,789 acres of the area, built the needed dike, and by the early 1940's the water level was rising and the ducks and teal were returning.

Just when the wildlife officials began to see success in their efforts at Montezuma, the engineers for the New York Thruway headed straight through the center of the refuge without saying anything of their plans. Indeed, the construction crews built the new superhighway to within sight of the boundary fence before anyone seemed to recognize what was on the drawing boards. The wildlife custodians tried hard to stand firm. But the Thruway engineers calmly pointed out that more than five million dollars' worth of finished road would have to be discarded to reroute the highway. To save New York state the five million, the federal government permitted the road to continue right through the refuge, destroying completely the only large lake on it. Today, Montezuma Refuge is reduced to perhaps a fourth of its former significance. How many of us think of this as we drive through its heart at sixty miles per hour and see only muskrat houses where vast flocks of waterfowl once rested and fed?

No longer is it the trapper or the hunter who most threatens wildlife. It is the government agency and the industrial corporation, eager to "reclaim" any large area not already producing a cash income. Sanctuaries, particularly on state or federal land, are especially vulnerable. One of the oldest, created in 1905 as the Wichita Mountains Game Preserve by executive order of President Theodore Roosevelt, has the misfortune to have as neighbor Fort Sill, Oklahoma, the U.S. Army's big artillery test center. The present Wichita Mountains National Wildlife Refuge, successor to the game preserve, harbors on its ninety-six square miles the largest herd of American bison in the country, as well as the few surviving longhorn cattle and a scattered wealth of elk, antelope, wild turkeys, and other game. It is the eastern extremity of range for the canyon wren and the roadrunner bird, and has become popular both as a place to see these and to fish.

To get space in which to test ever larger artillery pieces, the Army went directly to Congress, asking to have almost seventeen square miles at the south end of the refuge turned over to this military use. When this request was denied, a bill was slid through the legislative machinery, authorizing payment of the cost of transferring this generous sixth of the refuge to Fort Sill—without authorization for the transfer. The Secretary of the Interior managed to have this vetoed. One legislative maneuver after another was tried, and each was blocked in turn.

The Secretary yielded one concession: after 1955 the Army might haul into the southern end of the refuge some of its big artillery pieces and fire them back into Fort Sill. The effectiveness of areas within earshot as a haven for wild animals can be imagined. Animals vacated this part of the refuge to avoid the sudden invasion of tanks and halftracks, jeeps and trucks. But the Army continued to press for transfer of title, claiming that the testing of their equipment would be more adequate if the direction of fire were reversed, lobbing the shells into the refuge land from Fort Sill. The great value of the Wichita Refuge as a research center seems to mean nothing to the Army. Many of its officers continue to regard the refuge as an expendable federal zoo for native wildlife.

Dr. William Vogt, as Chief of the Conservation Section of the Pan American Union, spoke sadly in 1945 of "unsolved problems concerning wildlife in Mexican national parks." He pointed to a general apathy toward the fine efforts of Don Miguel de Quevado, who had succeeded in having a few important areas set aside for the future. Quevado regarded one of these as indispensable in conserving Mexico's always insufficient water supplies: the Colima National Park, which combined a fine forest, trapping moisture, with some of the country's greatest natural beauty and wildlife. Later the Mexican Congress changed the boundaries of the Colima Park to exclude almost the entire forested area, enabling a big corporation to cut the trees to make rayon. That the factory would close when the wood supply ended, and the water supply for the city of Colima would be jeopardized, bringing new problems to Mexico, disturbed neither the Congress nor the corporation.

We cannot criticize the Latin Americans without admitting to equivalent actions in our own country. On February 6, 1959, the selection of twelve hundred acres of federal land was announced for a new $110,000,000 "maximum security" federal prison. The Chief of the Bureau of Prisons had selected the site and recommended it to the House Appropriations subcommittee in closed session, to relieve overcrowding in federal prisons by providing for six hundred

more inmates. The site selected was on the shore of Crab
Orchard Lake and introduced only one complication. The
lake is in the middle of Crab Orchard National Wildlife
Refuge, an area set aside as a highly important way-station
for waterfowl migrating through Illinois and as an effective
breeding ground for ducks on the Mississippi Flyway. This
asset was to be destroyed for the sake of six hundred
criminals.

The way in which the economic motive conflicts with
conservation efforts has been particularly evident in Alaska
recently. Eighteen years ago, following the advice of hunters and wildlife scouts who studied the country from light
aircraft, a National Moose Range was set aside to protect a
breeding stock of the rangy, shaggy Alaskan moose, an animal standing seven feet tall and ten feet long. The 3,130
square miles of range were chosen in a marshy portion of
the Kenai Peninsula, across Cook Inlet from Anchorage.
Ten per cent of Alaska's moose were believed to live in this
1.8 per cent of Alaska's area. The marsh then had no known
commercial value.

Alaska is still too much of a frontier for anyone to think
of fencing the national moose range or of doing much to
patrol it. The area was marked on maps. Boundary signs
were erected. Yet the Richfield Oil Company entered the
federal property and drilled some test wells in the summer
of 1957. One of them came in with nine hundred barrels a
day, instantly bringing hopes of a northern land rush that
would rival the Klondike gold strike. Oilmen quickly filed
for leases on nearly six times the area of the national moose
range, inside its boundaries and all around it. Alaskans
sought to throw open the whole peninsula for exploration.
But the Department of the Interior, representing the U.S.
Fish and Wildlife Service, managers of the moose range,
fought for stiffer rules on granting oil leases—rules that

would apply to all U.S. game land. All leasing by the Department of the Interior was suspended until the new rules could be legally adopted.

In any conference on protection of wildlife, the unknowns are how many and what concessions will have to be made and whether any compensatory adjustments are possible. Some of the published comments during the argument on the national moose range make reading that is interesting in retrospect. Ex-governor Ernest Gruening, who argued in favor of unlimited exploitation of oil reserves regardless of moose, is quoted as stating that "the conservation lobbyists who get most upset are those who live in big-city apartments and have not the faintest practical notion of where the moose like to live." The December 16, 1957, issue of *Time* played down the importance of the moose range, pointing out that "in fact, only 10% of Alaska's moose live in the preserve." That the area was selected by experts, after a careful survey had shown that no other part of Alaska contains a comparable concentration of the big animals, went unmentioned.

So many wildlife sanctuaries have been obliterated or shifted to less suitable ground that they seem close parallels to *God's Little Acre* as described by Erskine Caldwell. Today wild animals often must depend upon man's leavings, if they are to survive. In Korea they find a sanctuary three miles wide in the demilitarized zone extending through Panmunjom across the country, since the area is forbidden territory to gun-carrying personnel other than the police guards of the Allied and Red Korean governments. Korean wildlife, remarkably varied for a country scarcely larger than Kansas, is thriving temporarily in the armistice area.

Australians are noticing that their famous koala "teddy bears" are easy to find in the outer suburbs of Brisbane, Melbourne, and Sidney, but rare in open country or most forests. Like the silver-gray opossum, the ring-tail opossum, the bandicoot, and many kinds of birds, the koalas tolerate close contact with human activities and benefit from the ban on use of firearms in populated places.

The ultimate in protection for wild creatures is surely a zoo. A number of kinds of animals can now be found nowhere else. The milou or Pere David's deer, *Elaphurus davidianus*, has been bred for centuries—perhaps from Marco Polo's time—in the Imperial Park at Pekin. The herd there was destroyed toward the end of the nineteenth century when war reached the area, but, fortunately, not before a few individuals had been sent to Europe. For a time the only existing herd was in Woburn Park, England. As these strange survivors, which shed their antlers twice yearly, became more numerous, breeding stock was shared with other zoos. In 1957 the *Proceedings of the Zoological Society of London* published a world-wide register of Pere David's deer, listing 380 individuals on several continents.

The European bison or wisent, *Bison bonasus*, finds sanctuary only in game parks and zoos. Since 1810 the center of its culture has been in Poland, under the sponsorship of an International Society for the Protection of the European Bison. This society had to suspend its operations during World War II, but resumed in 1944 only to learn that the once numerous surviving Polish wisents had shrunk to forty-one animals. Three years later a new studbook was published in Warsaw. It cited forty-four animals in Poland, twenty-eight in Germany, fifteen in Sweden, six in Russia, four each in England and the Netherlands, and one in Denmark. Within fifteen months the Polish herd had grown to fifty-three, and gamekeepers concluded that no degeneration had occurred. If man has sufficient interest in doing so, it appears that he can preserve the European bison indefinitely in captivity.

The status of the earth's sole remaining kind of wild horse is still in doubt. Recently Dr. Clement Purkyne of the

Zoological Gardens in Prague, Czechoslovakia, appealed to all owners of pure-bred Mongolian horses to send him details of age, sex, health, and breeding capacity so that he could compile a studbook of these animals. The stocky, short-maned Mongolian horse, *Equus przewalskii*, was named in honor of the Russian explorer Nikolai Przewalski, who discovered large herds of the animal on the plains of central Mongolia about the middle of the past century. Since that time the Mongolian horses have gone into a decline, partly through being hunted for meat, and partly by cross-breeding with domestic ponies. Whether any run wild in Mongolia today remains unknown.

At the turn of the century, after extended negotiations with the Tsar of Russia, the director of the zoo in Hamburg, Germany, went to Dzumgaria on the Mongolian plains and brought back thirty of Przewalski's horses to European fanciers. The animal has callosities on all four legs, a high-set tail, and seems intermediate between the domestic horse and wild asses. The descendants of these captive animals have been traded throughout the world, without much record to show present whereabouts. Until Dr. Purkyne's studbook has been compiled, it is impossible to guess whether the purity of this last wild species of horse can be maintained.

For a few kinds of wild animals, the future appears relatively secure because uncontrolled exploitation in the past seemed likely to exterminate them. Sanctuaries have been set aside in their behalf, operated under government monopolies. Today the Peruvians give utmost consideration to the guanay birds on offshore islands. The U.S. Fish and Wildlife Service supervises in minute detail the harvesting of each year's crop of fur seals in Alaska.

Off the Cape of Good Hope, the welfare of jackass penguins has become a responsibility of the provincial Department of Nature Conservation, to ensure that South African gourmets will have an unthreatened, although limited, supply of penguin eggs in season—at about twenty-five cents each. A penguin egg is approximately the size of a goose egg. After it has been hard-boiled, the white remains clear, the yolk is an exotic green, and the flavor so unusual that a taste for it must be cultivated.

Early and late in the long nesting period, the penguins are permitted to raise families undisturbed on isolated islands watched over from federal lighthouses. To visit the colonies at this time, we had to obtain special permission and to promise faithfully to show every consideration for the trusting penguins. In spring and summer, however, penguin eggs are collected systematically, with the proceeds from their sale paying for continued studies of the penguins' welfare.

Sanctuaries, however limited, require both initiative to establish and continuity over the years. Recently the trend has been toward small reserves to meet special needs. Perhaps the smallest wildlife refuge in the whole world was created in 1956 by the Netherlands government. The need for it arose as the Dutch people eliminated small bodies of standing water, regarding them as undesirable breeding places of mosquitoes. The change had disastrous effects on many amphibians, for now they had nowhere in which to congregate, lay eggs, and grow as tadpoles. The famous midwife toad *Alytes obstetricans* showed signs of disappearing through loss of sites in which to give "birth" to its offspring. As a sanctuary for midwife toads, a single pond—scarcely more than a big puddle—was created at the town of Epen. This miniature reserve seems to be attracting the amphibians and fulfilling all expectations.

Occasionally a whole town becomes a sanctuary for wild creatures. Many an American community proudly displays at its highway approaches the sign "Bird Sanctuary." Lesser creatures too are recognized. At Pacific Grove, California, a local ordinance protects the monarch butterflies that come in thousands every fall and cling to trees throughout the winter months. The trees selected by the butterflies are always the same ones, and these also are sacrosanct.

To some degree, each of us creates a wildlife sanctuary by attaching a bird-feeder to the window sill, putting out a bird bath, or building a garden pool. The beneficiaries may be only a few common kinds, yet their adoption of the improved living conditions we provide shows how little may be needed to tip the balance in favor of wild animals living near our homes.

We like to think that mankind is here to stay. Our presence on earth increasingly affects the survival and evolution of other kinds of life. But sanctuaries, no matter how well meant and tended, may merely postpone the day when the earth will have space only for creatures that adapt themselves to life at man's expense. No one has much good to say for the rat, the English sparrow, the European starling, the cockroach, or the housefly. Yet these are the animals that have altered their habits until products from our civilization serve them better than anything from their original homes. They compete with us successfully, and are in no danger of extinction. Their descendants may be our chief animal companions in the future unless man soon sees a great gain from letting life fend for itself on really large areas of the earth.

Every living thing also affects man's own evolutionary progress. When any kind of creature disappears, its influence on man goes too. The direction of his development surely changes slightly. Thoreau scoffed at people who believed they "could kill time without injuring eternity." By obliterating other kinds of life, man may be destroying himself as well.

Notes

The balance of nature has so many facets, each in such a fluid state, that the latest news tends to come in newspapers and magazines, often in columns written by reporters on the basis of press releases. After a time, progress reports appear in scientific journals, usually with a brief list of cited references to give the background. Later still, summary accounts become available in books; these can have more extensive bibliographies.

For readers wishing more information, the following special publications may prove helpful.

CHAPTER 1 · *Nature's Web*

P. 7, l. 34: H. S. Swingle: "Relationships and dynamics of balanced and unbalanced fish populations," *Alabama Experiment Station Bulletin* 274 (1950), pp. 1–74.

P. 9, l. 10: J. S. Dendy: "Bottom fauna in ponds with largemouth bass only and with a combination of largemouth bass plus bluegill," *Journal of the Tennessee Academy of Science*, Vol. xxxi, No. 3 (1956), pp. 198–207.

P. 9, l. 23: Swingle: ibid.

CHAPTER 2 · *A Land in Balance*

P. 13, l. 2: U.S. Department of Agriculture: *Grass: The Yearbook of Agriculture, 1948* (Washington, D.C.: U.S. Government Printing Office; 1949).

P. 14, l. 19: G. Catlin: *Illustrations of the Manners, Customs and Conditions of the North American Indians* (London: Henry G. Bohn; 1845).

P. 14, l. 23: A. H. Clark: "The impact of exotic invasion on the remaining New World mid-latitude grasslands," in *Man's Role in Changing the Face of the Earth*, edited by W. L. Thomas, Jr. (Chicago: University of Chicago Press; 1956) p. 742.

P. 15, l. 31: W. Gard: *The Great Buffalo Hunt* (New York: Alfred A. Knopf; 1959).

P. 16, l. 7: R. E. Smith: "Natural history of the prairie dog in Kansas," *University of Kansas Museum of Natural History and*

State Biological Survey, Miscellaneous Publications No. 16 (1958), pp. 1–36.

P. 20, l. 15: C. A. Fleschner: "Biological control of insect pests," *Science*, Vol. cxxix, No. 3348 (February 27, 1959), pp. 537–44.

P. 20, l. 21: C. A. Fleschner: "Insect pest control," *California Avocado Society Yearbook No. 39* [1954](1955), p. 155.

P. 21, l. 5: Smith: ibid.

CHAPTER 3 · Besting Nature

P. 24, l. 14: C. O. Sauer: "The agency of man on the earth," in *Man's Role in Changing the Face of the Earth*, edited by W. L. Thomas, Jr. (Chicago: University of Chicago Press; 1956), pp. 49–69.

P. 24, l. 19: C. S. Coon: *The Story of Man* (New York; Alfred A. Knopf; 1954).

P. 25, l. 26: E. K. Janaki Ammal: "Introduction to the subsistence economy of India," in *Man's Role . . .* (op. cit.), pp. 324–35.

P. 26, l. 13: I Kings, xv.

P. 26, l. 28: E. H. Graham: "The recreative power of plant communities," in *Man's Role . . .* (op. cit.), pp. 677–91.

P. 26, l. 32: E. Ayres: "The age of fossil fuels," in *Man's Role . . .* (op. cit.), pp. 367–81.

P. 30, l. 26: Y. Orev: "Brush invasion—1500 B.C. and 1950 A.D.," *Journal of Range Management*, Vol. ix, No. 1, pp. 6–7.

P. 33, l. 17: U.S. Department of Agriculture: *Food: The Yearbook of Agriculture, 1959* (Washington, D.C.: U.S. Government Printing Office; 1959).

P. 33, l. 19: Editorial, "Population explosion," *MD Medical Newsmagazine*, Vol. iii, No. 8 (August 1959), p. 61.

CHAPTER 5 · New Patches on Old

P. 43, l. 25: C. Swabey: "Plant introductions," *Glimpses of Jamaican Natural History*, Vol. i (1949), pp. 61–4.

P. 46, l. 18: C. B. Lewis: "Rats and the mongoose," *Glimpses of Jamaican Natural History*, Vol. i (1949), pp. 16–18; and "Rats and the mongoose in Jamaica," *Oryx*, Vol. ii, No. 3 (1953), pp. 170–2.

P. 46, l. 26: W. B. Espeut: "On the acclimatization of the Indian mungoos in Jamaica," *Proceedings of the Zoological Society of London, 1882*, pp. 712–14.

P. 47, l. 19: Espeut: ibid.

P. 48, l. 1: Espeut: ibid.

P. 48, l. 3: G. A. Seaman: "The mongoose and Caribbean wildlife," *Transactions of the 17th North American Wildlife Conference* (1952), pp. 188–97.

P. 49, l. 13: J. H. Westermann: "Nature preservation in the Caribbean," *Foundation for Scientific Research in Surinam and the Netherlands Antilles, Publication* 9 (1953), pp. 15–20.

P. 50, l. 26: G. N. Wolcott: "The rise and fall of the white grub in Puerto Rico," *American Naturalist*, Vol. LXXXIV, No. 816 (1950), pp. 183–93.

P. 51, l. 12: G. N. Wolcott: "The present status of economic entomology in Puerto Rico," *Transactions of the 9th International Congress of Entomology*, Vol. 1 (1952), pp. 776–7.

P. 52, l. 14: G. N. Wolcott: "The food of the mongoose (*Herpestes javanicus auropunctatus* Hodgson) in St. Croix and Puerto Rico," *Journal of Agriculture, University of Puerto Rico*, Vol. XXXVII, No. 3 (1953), pp. 241–7.

P. 53, l. 4: E. S. Tierkel *et al.:* "Mongoose rabies in Puerto Rico," *Public Health Reports*, Vol. LXVII, No. 3 (1952), pp. 274–8.

P. 53, l. 11: D. Pimentel: "The control of the mongoose in Puerto Rico," *American Journal of Tropical Medicine and Hygiene*, Vol. IV, No. 1 (1955), pp. 147–51.

P. 54, l. 1: P. H. Baldwin, C. W. Schwartz, and E. R. Schwartz: "Life history and economic status of the mongoose in Hawaii," *Journal of Mammalogy*, Vol. XXXIII, No. 3 (1952), pp. 335–56.

P. 54, l. 2: Seaman: ibid.

P. 54, l. 19: R. E. Smith: "Natural history of the prairie dog in Kansas," *University of Kansas Museum of Natural History and State Biological Survey, Miscellaneous Publications* No. 16 (1958), pp. 1–36.

CHAPTER 6 · *What Good Is a Crocodile?*

P. 57, l. 8: E. R. Allen and W. T. Neill: "The Florida crocodile," *Nature Magazine*, Vol. XLV, No. 2 (1952), pp. 77–80.

P. 59, l. 1: Editorial, "Where's Your Crocodile?", *The Laboratory* (Pittsburgh, Pa.), Vol. XXVI, No. 5 (1958), pp. 140–1.

P. 59, l. 11: C. B. Moore: "The grinning crocodile and his folklore," *Scientific Monthly*, Vol. LXXVIII, No. 4 (April 1954), pp. 225–31.

P. 60, l. 29: W. P. Bruce: "A preliminary test of dried crocodile meat in the feeding of pigs," *East African Agricultural Journal*, Vol. XV, No. 3 (1950), pp. 124–5.

P. 62, l. 34: Editorial, "Where's Your Crocodile?", loc. cit.

P. 63, l. 18: H. B. Cott: "The crocodile's decline," *The Listener*, Aug. 1, 1957, pp. 168–9.

P. 69, l. 23: M. Burton: "Exhibition of film of the fauna and flora of the Haute Semliki," *Proceedings of the Zoological Society of London*, Vol. CXXVI, No. 3 (1956), p. 488.

CHAPTER 7 · *A Yardstick for Deer*

P. 71, l. 7: E. C. O'Roke and F. N. Hamerstrom, Jr.: "Productivity and yield of the George Reserve deer herd," *Journal of Wildlife Management*, Vol. XII, No. 1 (1948), pp. 78–86.

P. 73, l. 16: G. H. Morton and E. L. Cheatum: "Regional differences in breeding potential of white-tailed deer in New York," *Journal of Wildlife Management,* Vol. x, No. 3 (1946), pp. 242–8.

P. 75, l. 15: Branch of Wildlife Research: "Big game inventory for 1956," *U.S. Department of the Interior, Fish and Wildlife Service, Wildlife Leaflet* No. 395 (1958), pp. 1–3.

P. 75, l. 18: U.S. Department of Agriculture: "Land utilization; a graphic summary, 1950," *1950 Census of Agriculture,* Vol. v, No. 4 (1952), pp. 1–58.

P. 76, l. 30: A. Starker Leopold: personal communication (1953).

P. 77, l. 16: M. H. Stenlund: "A field study of the timber wolf (*Canis lupus*) on the Superior National Forest, Minnesota," *Minnesota Department of Conservation Technical Bulletin* No. 4 (1955), pp. 1–55.

P. 78, l. 11: I. McT. Cowan: "The timber wolf in the Rocky Mountain national parks of Canada," *Canadian Journal of Research,* Section D, Zoological Sciences, Vol. xxv, No. 5 (1947), pp. 139–74.

P. 80, l. 9: R. P. Boone: "Deer management on the Kaibab," *Transactions of the 3rd North American Wildlife Conference* (1938), pp. 368–75.

P. 80, l. 22: D. I. Rasmussen: "Biotic communities of the Kaibab Plateau, Arizona," *Ecological Monographs,* Vol. xi (1941), pp. 229–75.

P. 82, l. 24: H. L. Schantz: "Big game populations in the national forests, 1921 to 1950," *Forest Science,* Vol. ii, No. 1 (1956), pp. 7–17.

P. 83, l. 13: L. W. Krefting: "Advancements in wildlife management on Indian lands," *Transactions of the 11th North American Wildlife Conference* (1946), pp. 434–41.

P. 83, l. 30: Anonymous: "National survey of fishing and hunting . . . during the calendar year 1955," *U.S. Department of Interior, Fish and Wildlife Service, Circular* 44 (1956), pp. 1–50.

P. 84, l. 3: L. B. Leopold, editor: *Round River: From the Journals of Aldo Leopold* (New York: Oxford University Press; 1953).

P. 84, l. 35: C. C. Presnall: "Changing trends in predator management," *Transactions of the 21st North American Wildlife Conference* (1956), pp. 185–90.

P. 85, l. 4: Krefting: ibid.

P. 85, l. 11: W. M. Longhurst and W. E. Howard: "Managing deer on private land" *California Agriculture,* Vol. x, No. 5 (1956), pp. 4, 10.

P. 85, l. 15: Longhurst and Howard: ibid.

P. 85, l. 22: E. B. Speaker: "Our deer, 1936–1952," *Iowa Conservationist,* Vol. xii, No. 1 (1953), pp. 97, 100.

P. 85, l. 24: J. R. Harlan: "The Iowa deer problem and its solution," *Iowa Conservationist*, Vol. xii, No. 1 (1953), pp. 97, 102.

P. 85, l. 29: Harlan: ibid.

P. 86, l. 3: W. E. Howard and W. M. Longhurst: "The farmer-sportsman problem and a solution," *Transactions of the 21st North American Wildlife Conference* (1956), pp. 323–30.

P. 86, l. 11: Longhurst and Howard: ibid.

P. 86, l. 17: Editorial, "Jackson County deer survey," *Wisconsin Conservation Bulletin*, Vol. xiii, No. 5 (1948), pp. 2–4.

P. 86, l. 33: S. G. de Boer: "Waste in the woods," *Wisconsin Conservation Bulletin*, Vol. xxii, No. 10 (1957), pp. 10–15.

P. 87, l. 13: Branch of Wildlife Research: ibid. Also Krefting: ibid.

P. 87, l. 25: K. Westerskov: "Observations on deer kill under different systems of hunting," *Journal of Wildlife Management*, Vol. xv, No. 1 (1951), pp. 27–32.

CHAPTER 8 · *The Big Bark-eaters*

P. 90, l. 6: A. Leopold: *A Sand County Almanac* (New York: Oxford University Press; 1949); p. 44.

P. 91, l. 10: J. Williams: *Fall of the Sparrow* (New York: Oxford University Press; 1951), p. 97.

P. 92, l. 30: E. T. Seton: *Lives of Game Animals* (New York: Doubleday, Doran and Co.; 1929).

P. 93, l. 15: L. E. Yeager and R. R. Hill: "Beaver management problems on western public lands," *Transactions of the 19th North American Wildlife Conference* (1954), pp. 462–80.

P. 94, l. 3: Williams: op. cit., p. 95.

P. 94, l. 24: W. Barker: *Familiar Animals of America* (New York: Harper and Bros.; 1956), pp. 69–70.

P. 97, l. 5: D. J. Neff: "Ecological effects of beaver habitat abandonment in the Colorado Rockies," *Journal of Wildlife Management*, Vol. xxi, No. 1 (1957), pp. 80–4.

P. 97, l. 12: W. S. Huey and W. H. Wolfrum: "Beaver-trout relations in New Mexico," *Progressive Fish-Culturist*, Vol. xviii, No. 2 (1956), pp. 70–4.

P. 99, l. 5: L. K. Couch: "Trapping and transplanting live beavers," *U.S. Department of the Interior, Fish and Wildlife Service, Conservation Bulletin* No. 30 (1942), pp. 1–20.

P. 99, l. 16: R. Twichell: "National status of beaver and beaver management," *Transactions of the 17th North American Wildlife Conference* (1952), pp. 476–82.

P. 99, l. 25: W. H. Lawrence, L. D. Fay, and S. A. Graham: "A report on the beaver die-off in Michigan," *Journal of Wildlife Management*, Vol. xx, No. 2 (1956), pp. 184–7.

P. 100, l. 12: Twichell: ibid.

P. 101, l. 9: B. Helmericks: "The lies they tell about wolverines!",

Field and Stream, Vol. LXI, No. 9 (1957), pp. 34–6, 67–8. Also P. Krott: *Demon of the North* (New York: Alfred A. Knopf; 1959). Also R. B. White: "The Wolverine," *Audubon Magazine*, Vol. LX, No. 2 (1958), pp. 66–7, 84.

P. 101, l. 30: F. G. Ashbrook: "Fur—an important wildlife crop," *Transactions of the 13th North American Wildlife Conference* (1948), pp. 465–74.

P. 103, l. 11: W. E. Dodge: personal communication (1957).

P. 103, l. 20: Dodge: ibid.

P. 103, l. 32: D. B. Cook and W. J. Hamilton, Jr.: "The forest, the fisher, and the porcupine," *Journal of Forestry*, Vol. LV, No. 10 (1957), pp. 719–22.

P. 104, l. 30: A. R. Shadle: "Gestation period of the porcupine," *Journal of Wildlife Management*, Vol. XXIX, No. 2 (1948), pp. 162–4. "Natural parturition of a porcupine and first reactions of the porcupette," *Ohio Journal of Science*, Vol. LIV, No. 1 (1954), pp. 42–4. "Effects of porcupine quills in humans," *American Naturalist*, Vol. LXXXIX, No. 844 (1955), pp. 47–9. Also "The North American porcupine up-to-date," *Natural Science Bulletin* (Ward's Natural Science Establishment), Vol. XXIV, No. 1 (1950), pp. 5–6, 11.

P. 105, l. 15: D. A. Spencer: "A forest mammal moves to the farm —the porcupine," *Transactions of the 11th North American Wildlife Conference* (1946), pp. 195–9. "Porcupines, rambling pincushions," *National Geographic Magazine*, Vol. XCVIII, No. 2 (August 1950), pp. 247–64. "The porcupine, its economic status and control," *U.S. Department of Interior, Fish and Wildlife Service, Wildlife Leaflet* No. 328 (1954), pp. 1–7.

P. 109, l. 5: W. H. Lawrence: "Porcupine control: a problem analysis," *Forestry Research Notes* (Weyerhaeuser Timber Company), pp. 1–43 (1957; mimeographed).

P. 109, l. 18: F. E. Beane: "Proposal to kill all N. H. Porcupines backed by agricultural advisors," Manchester (New Hampshire) *Union Leader*, Dec. 10, 1958.

P. 109, l. 18: W. E. Dodge: "An effective poison and repellent for porcupine control," *Journal of Forestry*, Vol. LVII, No. 5 (1959), pp. 350–2.

P. 109, l. 22: Anonymous: "Wisconsin turns to porcupine foe," *The New York Times*, December 7, 1958 [datelined Washington, November 29].

P. 109, l. 26: D. D. Berger: "Artificial nests for the great horned owl in Wisconsin," *Die Vogelwarte*, Vol. XVIII, No. 4 (1958), pp. 183–5.

P. 110, l. 3: G. T. Hamilton: "Resurgence of the fisher in New Hampshire," *Appalachia*, Vol. XXXI, No. 4 (1957), pp. 484–90. Concerning porcupines, by Lorus J. Milne and Margery Milne:

"Quillers are queer," *Canadian Nature*, Vol. XVIII, No. 5 (1957), pp. 162–5. "A prickly problem," *The New York Times Magazine,* March 15, 1959, pp. 97–8.

CHAPTER 9 · *Keeping a Closer Count*

P. 114, l. 3: Editorial, "Domestic pets," *Nature*, Vol. CLXXXI, No. 4617 (April 26, 1958), p. 1180.

P. 112, l. 16: R. Froman: "The buffalo that refused to vanish," *Reader's Digest*, Vol. LXXII, No. 434 (June 1958), pp. 144–54.

P. 117, l. 12: E. J. Kahn, Jr.: "Letter from Bermuda," *The New Yorker*, Vol. XXXII, No. 17 (June 16, 1956), pp. 117–20. Prat, H.: "Le genévrier dès Bermudes est-il menacé d'extinction?" *Bulletin de la Société botanique de France*, Vol. CII, No. 1–2 (1955), pp. 17–23. Also Waterson, J. M.: "The pests of juniper in Bermuda." *Tropical Agriculture* (Trinidad), Vol. XXVI, No. 1–6 (1949), pp. 5–15.

P. 122, l. 15: Editorial, "Wild Horse Annie," *Time*, July 27, 1959, p. 15. Also *Conservation Report* (National Wildlife Federation) No. 46 (Oct. 2, 1959), p. 347.

P. 123, l. 12: Editorial, "Rabbit 'Furbearer King' for 1956–57 Season," *South Dakota Conservation Digest*, Vol. XXIV, No. 6 (1958), pp. 6–7.

CHAPTER 10 · *New Weapons*

P. 129, l. 2: L. Crisler: *Arctic Wild* (New York: Harper and Bros.; 1958), p. 63.

P. 129, l. 25: Crisler: op. cit.

P. 130, l. 4: Crisler: op. cit.

P. 130, l. 14: L. J. Palmer and C. H. Rouse: "Study of the Alaska tundra with reference to its reactions to reindeer and other grazing," *U.S. Department of the Interior, Fish and Wildlife Service, Wildlife Report* No. 10 (1945), pp. 1–48.

P. 131, l. 13: Crisler: op. cit.

P. 131, l. 23: Crisler: op. cit.

P. 131, l. 32: F. F. Darling: "Man's ecological dominance through domesticated animals on wild lands," in *Man's Role in Changing the Face of the Earth*, edited by W. L. Thomas, Jr. (Chicago: University of Chicago Press; 1956), pp. 778–87.

P. 132, l. 8: V. B. Scheffer: "The rise and fall of a reindeer herd," *Scientific Monthly*, Vol. LXXIII, No. 6 (December 1951), pp. 356–62. Also F. Bourlière: *The Natural History of Mammals* (New York: Alfred A. Knopf; 1956), pp. 320–3.

P. 133, l. 4: A. Vincent: "Game Management," *Kenya Wild Life Society, 2nd Annual Report* (1957), p. 50.

P. 133, l. 17: C. W. Hobley: "Future of wild life in East Africa,"

Journal of the Society for the Preservation of the Fauna of the Empire, new series, Vol. LII (1945), pp. 30–3.

P. 133, l. 32: A. Vincent: "Poaching in Tanganyika," *Kenya Wild Life Society, 2nd Annual Report* (1957), p. 21. Also Hobley: op. cit., p. 33.

P. 134, l. 3: A. Vincent: "Poaching . . . ," op. cit., p. 19.

P. 134, l. 6: A. Vincent: "Poaching in Kenya," *Kenya Wild Life Society, 2nd Annual Report* (1957), p. 11.

P. 135, l. 1: A. Vincent: "Poaching in Tanganyika," op. cit., p. 18.

CHAPTER 11 · *Herdsmen's Pride*

P. 137, l. 1: R. W. Young, editor: *The Navajo Yearbook, Fiscal Year 1958* (Window Rock, Ariz.: The Navajo Agency; 1959), p. 318.

P. 138, l. 12: J. F. Dobie: *The Voice of the Coyote* (Boston: Little, Brown and Co.; 1949), pp. 43–4.

P. 139, l. 13: R. M. Underhill: *The Navajos* (Norman, Okla.: University of Oklahoma Press; 1956), p. 215.

P. 141, l. 10: F. F. Darling: "Man's ecological dominance through domesticated animals on wild lands," in *Man's Role in Changing the Face of the Earth* (Chicago: University of Chicago Press; 1956), p. 783.

P. 142, l. 3: A. Vincent: "The Serengeti National Park," *Kenya Wild Life Society, 2nd Annual Report* (1957), p. 26.

P. 143, l. 32: Vincent: ibid.

P. 144, l. 23: "Beacon" [pseudonym]: "Deserts waiting," reprinted from the *Kenya Weekly News* in *Kenya Wild Life Society, 2nd Annual Report* (1957), pp. 77–104; reference is to pp. 85–6.

P. 145, l. 5: "Beacon": ibid., p. 85.

P. 145, l. 22: "Beacon": ibid., p. 95.

P. 145, l. 28: A. Vincent: "Game management," *Kenya Wild Life Society, 2nd Annual Report* (1957), p. 50.

P. 145, l. 32: King George VI, quoted on letterhead of Kenya Wild Life Society, P.O. Box 20110, Nairobi, Kenya.

P. 146, l. 1: Proverbs, xvi, 18.

CHAPTER 12 · *Sequels to Spread*

P. 149, l. 12: P. S. Parker and R. E. Lennon: "Biology of the sea lamprey in its parasitic phase," *U.S. Department of the Interior, Fish and Wildlife Service, Research Report* No. 44 (1956), pp. 1–32.

P. 149, l. 34: V. C. Applegate: "The sea lamprey in the Great Lakes," *U.S. Department of the Interior, Fish and Wildlife Service, Leaflet* No. 384 (1950), pp. 1–8.

P. 150, l. 29: G. C. Pike: "Lamprey marks on whales," *Journal of the Fisheries Research Board of Canada*, Vol. VIII, No. 4 (1951), pp. 275–80.

P. 151, l. 26: C. L. Hubbs and T. E. B. Pope: "The spread of the sea lamprey through the Great Lakes," *Transactions of the American Fisheries Society*, Vol. LXVI (1937), pp. 172–6.

P. 151, l. 30: B. East: "Is the lake trout doomed?", *Natural History*, Vol. LVIII, No. 9 (1949), pp. 424–7.

P. 151, l. 32: D. S. Shetter: "A brief history of the sea lamprey problem in Michigan waters," *Transactions of the American Fisheries Society*, Vol. LXXVI (1949), pp. 160–76. Also R. Hile, P. H. Eschmeyer, and G. F. Lunger: "Decline of the lake trout fishery in Lake Michigan," *U.S. Department of the Interior, Fish and Wildlife Service, Bulletin* No. 52, Article No. 60 (1951), pp. 77–95.

P. 151, l. 33: F. E. J. Fry: "The 1944 class of lake trout in South Bay, Lake Huron," *Transactions of the American Fisheries Society*, Vol. LXXXII (1953), pp. 178–92.

P. 152, l. 2: Editorial, "Population changes and studies caused by the sea lamprey," *Progressive Fish-Culturist*, Vol. XVIII, No. 1 (1956), p. 41.

P. 153, l. 14: J. L. Thomas: "The life cycle of *Diphyllobothrium oblongatum* Thomas, a tapeworm of gulls," *Journal of Parasitology*, Vol. XXXIII, No. 2 (1947), pp. 107–17.

P. 154, l. 6: J. W. Moffett: "Attack on the sea lamprey," *Michigan Conservation Magazine*, May–June 1958.

P. 154, l. 27: V. C. Applegate, B. R. Smith, and W. L. Nielsen: "Use of electricity in the control of sea lampreys: electromechanical weirs and traps and electrical barriers," *U.S. Department of the Interior, Fish and Wildlife Service, Special Scientific Report—Fisheries*, No. 92 (1952), pp. 1–52.

P. 155, l. 13: E. J. McVeigh: "Showdown in the Great Lakes," *The Dow Diamond* (Dow Chemical Co.), June 1958, pp. 1–8.

P. 155, l. 23: V. C. Applegate, J. H. Howell, and M. A. Smith: "Use of mononitrophenols containing halogens as selective lamprey larvicides," *Science*, new series, Vol. CXXVII, No. 3294 (1958), pp. 336–8.

P. 155, l. 29: D. Stetson: "Trout to be put in Lake Superior," *The New York Times*, November 15, 1959, p. 62.

P. 156, l. 19: P. B. van Weel: "Some notes on the African giant snail, *Achatina fulica* Fer. III. Observations on its biology," *Chronica Natura*, Vol. CIV, No. 12 (1948), pp. 335–6.

P. 157, l. 19: F. X. Williams: "Life history studies of East African *Achatina* snails," *Bulletin of the Museum of Comparative Zoölogy, Harvard University*, Vol. CV, No. 3 (1951), pp. 293–317.

P. 157, l. 30: A. R. Mead: "The giant snails," *Atlantic Monthly*, Vol. CLXXXIV, No. 2 (1949), pp. 38–42.

P. 159, l. 6: S. Hatai and K. Genji: [Observations upon growth of shells and ecology of *Achatina fulica* Ferussac in Palau], *Kagaku Nanyo* [*Science of the South Seas*], Vol. V, No. 2 (1943), pp. 1–19 [in Japanese].

P. 159, l. 9: T. Esaki and K. Takahashi: "Introduction of the African snail *Achatina fulica* Ferussac into Japan, especially Micronesia," *Kagaku Nanyo*, Vol. IV, No. 3 (1942), pp. 16–25.

P. 160, l. 7: A. R. Mead: "The proposed introduction of predatory snails into California," *The Nautilus*, Vol. LXIX, No. 2 (1955), pp. 37–40.

P. 162, l. 10: R. Lewinsohn: *Animals, Men and Myths* (New York: Harper and Bros.; 1954), p. 315.

P. 162, l. 13: E. W. Stearn and A. E. Stearn: *The Effect of Smallpox on the Destiny of the Amerindian* (Boston: Humphries; 1945).

P. 162, l. 21: M. Bates: "The natural history of yellow fever in Columbia," *Scientific Monthly*, Vol. LXIII, No. 1 (1946), pp. 42–52.

P. 162, l. 25: E. C. Faust: "The history of malaria in the United States," *American Scientist*, Vol. XXXIX, No. 1 (1951), pp. 121–8.

CHAPTER 13 · *Conquering Rabbits*

P. 164, l. 4: R. T. Peterson: "The great natural experiment," *Audubon Magazine*, Vol. LVIII, No. 6 (1956), pp. 248, 250–1, 287.

P. 165, l. 1: J. F. Hart: "Myxomatosis in Britain," *Geographic Review*, Vol. XLVII, No. 1 (1957), pp. 126–8.

P. 165, l. 15: D. Allen: *Our Wildlife Legacy* (New York: Funk and Wagnalls; 1954).

P. 165, l. 34: K. A. Wodzicki: *Introduced Mammals of New Zealand* (Wellington, New Zealand: Department of Scientific and Industrial Research, Bulletin No. 98; 1950).

P. 166, l. 6: A. D. McIntosh: *Marlborough, A Provincial History* (Blenheim; 1940).

P. 166, l. 18: J. Williams: *Fall of the Sparrow* (New York: Oxford University Press; 1951).

P. 167, l. 12: E. Troughton: *Furred Animals of Australia* (New York: Charles Scribner's Sons; 1947).

P. 167, l. 17: Wodzicki: op. cit.

P. 168, l. 26: C. M. Herman: "A review of experiments in biological control of rabbits in Australia," *Journal of Wildlife Management*, Vol. XVII, No. 4 (1953), pp. 482–6.

P. 169, l. 10: J. E. Nichols: "Rabbit control in Australia: problems and possibilities," *Nature*, Vol. CLXVIII, No. 4283 (1951), pp. 932–4.

P. 169, l. 23: J. LeG. Brereton: "Initial spread of myxomatosis in Australia," *Nature*, Vol. CLXXII, No. 4368 (1953), pp. 108–10.

P. 170, l. 3: *Time*, August 10, 1953.

P. 171, l. 13: J. E. Montel: "Rabbit production in Italy," *American Rabbit Journal*, Vol. XXI, No. 8 (1951), p. 118.

P. 171, l. 22: J. A. Valverde: "Notes ecologiques sur le lynx d'Espagne, *Felis lynx pardina* Temminck," *La Terre et la Vie* (Paris), Vol. CIV, No. 1 (1957), pp. 51–67.

P. 171, l. 34: H. V. Thompson: "The rabbit disease: myxomatosis," *Annals of Applied Biology*, Volume XLI, No. 2 (1954), pp. 358–66.

P. 172, l. 33: J. N. Ritchie, J. R. Hudson, and H. V. Thompson: "Myxomatosis," *Veterinary Record*, Volume LXVI (1954), pp. 796–804.

P. 172, l. 9: *The New York Times*, May 1, 1955.

P. 172, l. 11: *Science*, new series, Vol. CXXII, No. 3176 (November 11, 1955), p. 915.

P. 172, l. 12: G. Cobnut: "Foxes and myxomatosis in Kent, *Oryx* (London), Vol. III (1955), pp. 156–7.

P. 173, l. 15: F. N. Ratcliffe *et al.*: "Myxomatosis in Australia, a step toward the biological control of the rabbit," *Nature*, Vol. CLXX, No. 4314 (1952), pp. 7–11.

P. 173, l. 21: D. J. Lee, K. J. Clinton, and A. K. O'Gower: "The blood sources of some Australian mosquitoes," *Australian Journal of Biological Sciences*, Vol. VII, No. 3 (1954), pp. 282–301.

P. 174, l. 10: H. V. Thompson: "Myxomatosis: recent developments," *Journal of the Ministry of Agriculture* [Australia], Vol. LXI (1955), pp. 317–21.

P. 175, l. 1: R. M. Licksley: "Failure of myxomatosis on Skokholm Island," *Nature*, Vol. CLXXV, No. 4464 (1955), pp. 906–7.

P. 175, l. 11: M. Rothschild and H. Marsh: "Increase of hares (*Lepus europaeus* Pallas) at Ashton Wold, with a note on the reduction in numbers of the brown rat (*Rattus norvegicus* Berkenhout)," *Proceedings of the Zoological Society of London*, Vol. CXXVII, No. 3 (1956), pp. 441–5.

P. 175, l. 17: H. N. Southern: "Ecologists are excited by England's 'rabbit disease,'" *Animal Kingdom*, Vol. LIX, No. 4 (1956), pp. 116–23.

P. 175, l. 28: Peterson: op. cit.

P. 175, l. 35: Peterson: op. cit.

P. 176, l. 20: Editorial, "Myxomatosis in Australia," *Nature*, Vol. CLXXVIII, No. 4544 (1956), p. 1219.

P. 177, l. 9: B. V. Fennessy: "The control of the European rabbit in New Zealand," *Commonwealth Scientific and Industrial Research Organization, Wildlife Survey Section, Technical Paper* No. 1 (1958), pp. 1–40.

P. 177, l. 26: W. E. Howard: "Rabbit control in New Zealand," *California Livestock News*. Vol. xxxiv, No. 25 (1958), p. 14.

P. 178, l. 2: W. E. Howard: "The rabbit problem in New Zealand," *New Zealand Department of Science and Industrial Research, Information Series*, No. 16 (1958), pp. 1–47.

P. 178, l. 17: H. V. Thompson: "Myxomatosis: recent developments," *Journal of the Ministry of Agriculture*, Vol. lxi (1955), pp. 317–21.

P. 178, l. 20: H. V. Thompson: "The wild European rabbit and possible dangers of its introduction into the U.S.A.," *Journal of Wildlife Management*, Vol. xix, No. 1 (1955), pp. 8–13.

P. 179, l. 9: P. L. Shanks *et al.*: "Experiments with myxomatosis in the Hebrides," *British Veterinary Journal*, Vol. cxi, No. 1 (1955), pp. 25–30.

P. 179, l. 22: R. M. Latham: "The controversial San Juan rabbit," *Transactions of the 20th North American Wildlife Conference* (1955), pp. 406–14.

P. 180, l. 24: K. Bednarik: "San Juan rabbits: furred menace?", *Ohio Conservation Bulletin*, Vol. xix, No. 12 (1955), pp. 8–9, 31–2.

P. 180, l. 32: L. S. Rudasill: "An experiment with the San Juan rabbit," *Maryland Conservationist*, Vol. xxxiii, No. 5 (1956), pp. 8–9.

P. 180, l. 34: W. B. Barnes: "The status of the San Juan rabbit in the Midwest," *Transactions of the 17th Midwest Wildlife Conference* (1955), pp. 1–7.

P. 180, l. 35: *Science*, new series, Vol. cxx, No. 3131 (December 31, 1954), p. 1087.

P. 181, l. 2: Latham: op. cit.

P. 181, l. 6: H. V. Thompson: "The wild European rabbit . . ." (1955), op. cit.

CHAPTER 14 · *Useful Wetlands*

P. 182, l. 16: F. Schiemenz: "Die Umänderung der sich wandelnden Flüsse in feste Gebilde und die Auswirkung auf die Fischerei," *Fischerei-Zeitung, Kriegsgemeinschaftsausgabe* xlvii, No. 3–4 (1944), pp. 9–11.

P. 183, l. 4: A. M. Day: "The problem of increased hunting pressure on waterfowl," *Transactions of the 11th North American Wildlife Conference* 1946, pp. 55–61.

P. 183, l. 11: S. P. Shaw and C. G. Fredine: "Wetlands of the United States," *U.S. Department of Interior, Fish and Wildlife Service, Circular* No. 39 (1956), pp. 1–67.

P. 183, l. 21: C. Cottam and W. S. Bourn: "Coastal marshes adversely affected by drainage," *U.S. Department of the Interior, Fish and Wildlife Service*, illustrated lecture (1952), pp. 1–7 [mimeographed].

P. 184, l. 6: R. W. Schery: *Plants for Man* (New York: Prentice-Hall; 1952).

P. 184, l. 18: A. E. Hofstede and R. O. Ardiwinata: "Compiling statistical data on fish culture in irrigated rice fields in west Java," *Landbouw*, Vol. xxii, No. 10–12 (1950), pp. 469–94.

P. 184, l. 23: D. W. LeMara: " 'Weeding' in fish farming," *Nature*, Vol. clxii, No. 4122 (1948), p. 704.

P. 184, l. 31: W. F. Sigler: "Carp as a protein supplement," *Farm and Home Science*, Vol. x, No. 2 (1949), pp. 10–11.

P. 185, l. 9: W. Wunder: "Verschiedenartige Nutzung von Karpfenteichen," *Allgemeine Fischereizeitung*, Vol. lxxii, No. 23–24 (1947), pp. 300–6.

P. 185, l. 22: W. Wunder: "Karpfenteichwirtschaft in Ungarn," *Allgemeine Fischereizeitung*, Vol. lxxv, No. 5 (1950), pp. 123–4.

P. 185, l. 33: J. S. Grim: "The carp," *New York State Conservationist*, Vol. v, No. 1 (1950), pp. 10–12.

P. 186, l. 12: C. W. Threinen and W. T. Helm: "Experiments and observations designed to show carp destruction of aquatic vegetation," *Journal of Wildlife Management*, Vol. xviii, No. 2 (1954), pp. 247–51.

P. 186, l. 26: M. L. Giltz and W. C. Myser: "A preliminary report on an experiment to prevent cattail die-off," *Ecology*, Vol. xxxv, No. 3 (1954), p. 418.

P. 187, l. 3: D. Mraz and E. L. Cooper: "Natural reproduction and survival of carp in small ponds," *Journal of Wildlife Management*, Vol. xxi, No. 1 (1957), pp. 66–9.

P. 187, l. 30: A. D. Holloway: "Twelve years of fishing records from Lake Mattamuskeet," *Transactions of the 13th North American Wildlife Conference* (1948), pp. 474–80.

P. 188, l. 22: W. G. Cahoon: "Commercial carp removal at Lake Mattamuskeet, North Carolina," *Journal of Wildlife Management*, Vol. xvii, No. 3 (1953), pp. 312–17.

P. 191, l. 3: J. L. Farley: "Duck stamps and wildlife refuges," *U.S. Department of the Interior, Fish and Wildlife Service, Circular* No. 37 (1955), pp. 1–22.

P. 141, l. 7: Farley: ibid.

P. 192, l. 30: S. G. Jewett: "Klamath Basin National Wildlife Refuges, Oregon and California," *U.S. Department of the Interior, Fish and Wildlife Service, Leaflet* No. 238 (1943), pp. 1–18 [mimeographed].

P. 193, l. 17: Editorial, "Will wildlife get water at Klamath-Tule area?", *Conservation News*, Vol. xxiv, No. 22 (1959), pp. 3–5.

P. 194, l. 25: C. E. Renn: "The wasting disease of common eelgrass, *Zostera marina*," *Biological Bulletin*, Vol. lxx (1936), pp. 148–57.

P. 195, l. 7: P. Manfredi: "[Some products of the sea]" *Natura* (Milan), Vol. xxxiii, No. 3–4 (1942), pp. 106–10.

P. 195, l. 16: Editorial note regarding decline of jellyfish *Gorionemus* near Woods Hole, Massachusetts, *Turtox News* (General Biological Supply House), Vol. xxxvii, No. 3 (March 1959), p. 85.

P. 195, l. 30: D. P. Wilson: "The decline of *Zostera marina* L. at Salcombe and its effect on the shore," *Journal of the Marine Biological Association of the United Kingdom,* Vol. xxviii, No. 2 (1949), pp. 395–412.

P. 196, l. 23: C. Cottam and D. A. Munro: "Eelgrass status and environmental relations," *Journal of Wildlife Management,* Vol. xviii, No. 4 (1954), pp. 449–60.

P. 197, l. 9: U.S. Department of Agriculture: *Climate and Man: Yearbook of Agriculture, 1941* (Washington, D.C.: U.S. Government Printing Office; 1941).

CHAPTER 15 · *Waste Not, Want Not*

P. 199, l. 23: P. S. Galtsoff: "Ecological changes affecting the productivity of oyster grounds," *Transactions of the 21st North American Wildlife Conference* (1956), pp. 408–19.

P. 202, l. 29: D. C. Hogner: *Conservation in America* (Philadelphia: Lippincott; 1958), p. 215.

P. 202, l. 33: K. E. Mundt: "Pollution a measure of civilization," *Wisconsin Conservation Bulletin,* Vol. viii, No. 1 (1948), pp. 14–15.

P. 203, l. 6: W. E. Towell: "Freshwater cancer," *Missouri Conservationist,* Vol. xix, No. 9 (1958), pp. 1–3.

P. 204, l. 8: D. Allen: *Our Wildlife Legacy* (New York: Funk and Wagnalls; 1954), p. 76.

P. 204, l. 27: W. Vogt, Jr.: "Does pollution go on forever?", *Transactions of the 13th North American Wildlife Conference* (1948), pp. 129–35.

P. 204, l. 29: C. E. ZoBell: personal communication.

P. 205, l. 30: D. Merriman: "El Niño brings rain to Peru," *American Scientist,* Vol. xliii, No. 1 (1955), pp. 63–76.

P. 206, l. 28: H. U. Sverdrup, M. W. Johnson, and R. H. Fleming: *The Oceans: Their Physics, Chemistry, and General Biology.* (New York: Prentice-Hall; 1942).

P. 207, l. 33: C. R. Odin: "California gull predation on waterfowl," *The Auk,* Vol. lxxiv, No. 2 (1957), pp. 185–202.

P. 208, l. 6: A. C. Twomey: "California gulls and exotic eggs," *The Condor,* Vol. l, No. 3 (1948), pp. 97–100.

P. 208, l. 26: Anonymous, reprinted in *Audubon Magazine,* Vol. lx, No. 5 (September–October 1958), p. 220, from *The New York Times* for July 15, 1958.

P. 208, l. 33: A. O. Gross: "Gulls of Muskeget Island," *Bulletin of the Massachusetts Audubon Society,* Vol. xxxii, No. 2 (1948), pp. 43–7.

P. 209, l. 15: Otterlind, G.: [Nutrition and distributional ecology of the herring gull *Larus a. argentatus* Pont.] *Kongliga Fysiografiska Sällskapet i Lund Förhandlingar*, Vol. XVIII (1948), pp. 36–56.

P. 210, l. 18: M. F. Morzer Bruijns: [The herring gull problem in the year 1956] *Levende Natuur*, Vol. LIX, No. 6 (1956), pp. 128–31.

P. 210, l. 21: A. O. Gross: "The herring gull-cormorant control project, 1952," *U.S. Department of the Interior, Fish and Wildlife Service, Report for Region 5* (1952), pp. 1–14.

P. 211, l. 16: J. A. Livingston: "The senseless slaughter of our sea-birds," *Maclean's Magazine*, August 2, 1958, pp. 24, 30–1.

P. 212, l. 8: Livingston: ibid.

P. 212, l. 19: Livingston: ibid.

CHAPTER 16 · *Tree Country*

P. 215, l. 28: R. E. McArdle: "Timber resources for America's future," An address, summarizing the *Timber Resource Review*, before the annual meeting of the Society of American Foresters, Portland, Oregon, October 17, 1955.

P. 218, l. 3: U.S. Department of Agriculture: *Trees: Yearbook of Agriculture, 1949* (Washington, D.C.: U.S. Government Printing Office; 1949), p. 472.

P. 218, l. 16: C. Darwin: *Journal of Researches into the Geology and Natural History of the Various Countries Visited during the Voyage of H.M.S. Beagle Round the World* (London: 1839).

P. 218, l. 31: H. L. Stoddard: *The Bobwhite Quail* (New York: Charles Scribner's Sons; 1936).

P. 219, l. 21: C. E. Randall: "Forest Service, U.S.," in *New International Yearbook, 1958* (New York: Funk and Wagnalls; 1959).

P. 219, l. 31: *Time*, July 20, 1959, p. 22.

P. 220, l. 2: *Time*, ibid.

P. 221, l. 8: A. H. Carhart: *Trees and Game—Twin Crops* (Washington, D.C.: American Forest Products Industries).

P. 222, l. 7: Carhart: ibid.

P. 222, l. 14: U.S. Department of Agriculture: "People and Timber," *U.S.D.A., Forest Service, Miscellaneous Publication No. 721* (1956), pp. 1–16.

P. 222, l. 18: L. F. Watts: "Insects and diseases—the greatest forest menace," *Transactions of the 17th North American Wildlife Conference* (1952), pp. 20–6.

P. 222, l. 20: Carhart: op. cit.

P. 222, l. 25: P. B. Sears: "Wildlife in today's economy: biological and ecological values," *Transactions of the 16th North American Wildlife Conference* (1951), pp. 23–6.

P. 222, l. 27: N. M. Kurennoy and G. B. Ankinovich: [Bees and the seed qualities of acorns] *Pchelovodstvo*, Vol. xxxiii, No. 9 (1956), pp. 43–6 [in Russian].

P. 223, l. 19: R. S. Hyslop: "Moose and spruce," *Sylva*, Vol. xiv, No. 4 (1958), pp. 40–1.

P. 228, l. 20: H. P. H. Behrens: "Bush-fence barrier," *African Wild Life*, Vol. i, No. 2 (1947), pp. 15–19.

P. 229, l. 28: J. P. Glasgow and F. Wilson: "A census of the tsetse fly *Glossina pallidipes* Austin and of its host animals," *Journal of Animal Ecology*, Vol. xxii, No. 1 (1953), pp. 47–56.

P. 230, l. 15: A. Esteves de Sousa: "Preliminary experiments with phytocides to control stump growth and thorny bush thickets in areas infested with *Glossina austeni*," *Moçambique*, Vol. lxxxvii (1956), pp. 83–110.

P. 230, l. 29: W. H. Potts and C. H. N. Jackson: "The Shinyanga game destruction experiment," *Bulletin of Entomological Research*, Vol. xliii, No. 2 (1952), pp. 365–74.

P. 231, l. 13: Trypanosomiasis Committee of Southern Rhodesia: "The scientific basis of the control of *Glossina morsitans* by game destruction," *Journal of the Society for the Preservation of the Fauna of the Empire*, new series, Vol. liv (1946), pp. 10–16.

P. 231, l. 15: D. L. Thrapp: "S.O.S. for African wildlife," *Natural History*, Vol. lviii, No. 3 (1949), pp. 104–11, 141–2.

P. 231, l. 19: J. Stevenson-Hamilton: "A game warden reflects," *Journal of the Society for the Preservation of the Fauna of the Empire*, new series, Vol. liv (1946), pp. 17–21.

P. 231, l. 35: Anonymous: "Reprieve for wild animals," *Journal of the Society for the Preservation of the Fauna of the Empire*, new series, Vol. lvii (1948), pp. 30–4.

P. 232, l. 13: P. B. Sears: "An ecological view of land-use in Middle America," *Ceiba*, Vol. iii, No. 3 (1953), pp. 157–65.

P. 233, l. 20: P. B. Sears: "Wildlife in today's economy . . . ," op. cit. (1951).

CHAPTER 17 · *"The Greatest Threat to Life on Earth"*

P. 236, l. 16: H. S. Satterlee: "The problem of arsenic in American cigarette tobacco," *New England Journal of Medicine*, Vol. ccliv (1956), pp. 1149–54.

P. 236, l. 23: Satterlee: ibid.

P. 236, l. 29: O. Warburg: "On the origin of cancer cells," *Science*, new series, Vol. cxxiii, No. 3191 (February 24, 1956), pp. 309–14.

P. 237, l. 1: Satterlee: ibid.

P. 237, l. 7: *Time*, August 30, 1959, p. 71.

P. 237, l. 31: A. W. A. Brown: "The spread of insecticide resistance in pest species," in *Advances in Pest Research*, edited

NOTES

by R. L. Metcalf (New York: Interscience Publishers), Vol. II (1957), pp. 351–97.

P. 238, l. 16: H. Zinsser: *Rats, Lice and History* (Boston: Little, Brown; 1934).

P. 238, l. 35: A. H. Benton and W. E. Werner, Jr.: *Principles of Field Biology and Ecology* (New York: McGraw-Hill Book Co.; 1958), p. 161.

P. 239, l. 1: Brown: op. cit.

P. 239, l. 4: Brown: op. cit.

P. 239, l. 14: Brown: op. cit.

P. 240, l. 26: *Conservation News*, Vol. XXIV, No. 18 (September 15, 1959).

P. 241, l. 26: Benton and Werner: op. cit.

P. 242, l. 6: P. F. Springer: "Insecticides: boon or bane?" *Audubon Magazine*, Vol. LVIII, No. 4 (1956), pp. 176–8.

P. 242, l. 14: J. B. de Witt: "Effects of chemical sprays on wildlife," *Audubon Magazine*, Vol. LX, No. 2 (1958), pp. 70–1.

P. 242, l. 17: C. S. Robbins *et al.*: "Effects of five-year DDT application on a breeding bird population," *Journal of Wildlife Management*, Vol. XV, No. 2 (1951), pp. 213–16.

P. 242, l. 20: C. S. Robbins and R. E. Stewart: "Effects of DDT on bird population of scrub forest," *Journal of Wildlife Management*, Vol. XIII, No. 1 (1949), pp. 11–16.

P. 242, l. 22: A. H. Benton: "Effects on wildlife of DDT used for control of Dutch elm diseases," *Journal of Wildlife Management*, Vol. XV, No. 1 (1951), pp. 20–7.

P. 242, l. 27: H. P. Blagbrough: "Reducing wildlife hazards in Dutch elm disease control," *Journal of Forestry*, Vol. L, No. 6 (1952), pp. 468–9.

P. 242, l. 35: G. J. Wallace: "Insecticides and birds," *Audubon Magazine*, Vol. LXI, No. 1 (1959), pp. 10–12, 35. "Another year of robin losses on a university campus," ibid., Vol. LXII, No. 2 (1960) pp. 66–9.

P. 243, l. 4: R. J. Barker: "Notes on some ecological effects of DDT sprayed on elms," *Journal of Wildlife Management*, Vol. XXII, No. 2 (1958), pp. 269–74.

P. 243, l. 13: J. B. de Witt: "Birds and Dutch elm disease control," *Proceedings of the 13th Midwest Shade Tree Conference* (1958), pp. 3–7.

P. 243, l. 20: D. A. Greenwood *et al.*: "Feeding rats tissues from lambs and butterfat from cows that consumed DDT-dusted alfalfa hay," *Proceedings of the Society for Experimental Biology and Medicine*, Vol. LXXXIII, No. 3 (1953), pp. 458–60.

P. 243, l. 28: J. B. Oakes: "Conservation: nature's balance," *The New York Times*, June 2, 1957.

P. 244, l. 4: Editorial, "Court upholds DDT spraying," *American Forestry*, Vol. LXIV, No. 8 (1958), pp. 23, 55–6.

P. 244, l. 10: "The hick mind in action?", *American Forestry*, Vol. LXIV, No. 2 (1958), pp. 4, 51–2.

P. 244, l. 13: J. H. Baker: "The greatest threat to life on earth," *Outdoor America*, June 1958, pp. 4–5.

P. 245, l. 3: Baker: ibid.

P. 245, l. 7: *Time*, December 29, 1958, p. 12. Also *Consumer Reports*, October 1959 and January 1960.

P. 245, l. 19: *The New Yorker*, December 12, 1959, p. 187.

P. 245, l. 24: *The New York Times*, November 22, 1959.

P. 245, l. 30: *The New Yorker*, op. cit.

P. 246, l. 3: P. R. Nelson: "Effects of fertilizing Bare Lake, Alaska, on growth and production of red salmon (*O. nerka*)," *U.S. Department of the Interior, Fish and Wildlife Service, Fishery Bulletin* No. 60 (1959), pp. 59–86.

P. 246, l. 8: F. P. Ide: "Effect of forest spraying with DDT on aquatic insects of salmon streams," *Transactions of the American Fisheries Society*, Vol. LXXXVI (1957), pp. 208–19.

P. 246, l. 18: C. Lyle and I. Fortune: "Notes on an imported fire ant," *Journal of Economic Entomology*, Vol. XLI, No. 5 (1948), pp. 833–4.

P. 246, l. 28: H. W. Bates: *The Naturalist on the Amazons* (London: J. Murray; 1863).

P. 247, l. 4: *Life*, Vol. XLV, No. 2 (July 14, 1958), pp. 117–18.

P. 247, l. 34: W. G. Eden and F. S. Arant: "Control of the imported fire ant in Alabama," *Journal of Economic Entomology*, Vol. XLII, No. 6 (1950), pp. 976–9.

P. 248, l. 2: U.S. Department of Agriculture: *Insects: Yearbook of Agriculture, 1953* (Washington, D.C.: U.S. Government Printing Office; 1953).

P. 248, l. 12: H. S. Peters: "Late news from the fire ant front," an address before the 55th Annual Convention of the National Audubon Society, November 10, 1959 [one-page mimeographed news release].

P. 248, l. 25: M. Whisenhunt: "The fire ant eradication program and wildlife in Florida," *Florida Wildlife*, Vol. XII, No. 6 (1958), p. 5.

P. 248, l. 30: R. H. Stroud: "Super insecticides—space age pollutants," *Sport Fishing Institute Bulletin*, Vol. LXXIV (1958), pp. 1–2.

P. 249, l. 1: Editorial, "The greatest killing program of all?", *Audubon Magazine*, Vol. LX, No. 6 (1958), pp. 254–5, 294. Also D. W. Lay: "Count three for trouble," *Texas Game and Fish*, Vol. XVI, No. 7 (1958), pp. 4–7.

P. 249, l. 6: M. F. Baker: "Observations on effects of an application of heptachlor or dieldrin on wildlife," *Proceedings of the 12th Annual Conference of the Southeastern Association of Game and Fish Commissions 1958* (1959), [preprint pp. 18–21]. Also S. G. Clawson: "Fire ant eradication . . . and quail," *Alabama Conservationist*, Vol. xxx, No. 4 (1959), pp. 14–15, 25. Also S. G. Clawson and M. F. Baker: "Immediate effects of dieldrin and heptachlor on bobwhites," *Journal of Wildlife Management*, Vol. xxiii, No. 2 (1959), pp. 215–19. Also D. W. Lay: "Fire ant eradication and wildlife," *Proceedings of the 12th Annual Conference of the Southeastern Association of Game and Fish Commissions 1958* (1959), [preprint pp. 22–4].

P. 249, l. 21: H. S. Peters: "Is it possible you are being stung?", *Florida Wildlife*, Vol. xii, No. 6 (1958), pp. 4, 42.

P. 249, l. 27: L. L. Glasgow: "Studies on the effect of the imported fire ant control program on wildlife in Louisiana," *Proceedings of the 12th Annual Conference of the Southeastern Association of Game and Fish Commissions 1958* (1959) [preprint pp. 24–9]. Also J. D. Newson: "A preliminary report on fire ant eradication program, Concordia Parish, Louisiana, June, 1958," ibid. [preprint pp. 29–31].

P. 250, l. 1: Editorial, "The greatest killing program of all?", *Audubon Magazine*, Vol. lx, No. 6 (1958), pp. 254–5, 294.

P. 250, l. 11: U.S. Department of Agriculture: *Soil: Yearbook of Agriculture, 1957* (Washington, D.C.: U.S. Government Printing Office; 1957).

P. 250, l. 14: H. S. Peters: "Late news from the fire ant front," op. cit.

P. 250, l. 23: *Conservation News*, Vol. xxiv, No. 21 (November 1, 1959), pp. 1–2.

P. 250, l. 27: C. E. Broley: "The bald eagle in Florida," *Atlantic Naturalist*, Vol. xii, No. 5 (1957), pp. 230–1.

P. 250, l. 33: J. B. de Witt: "Effects of chemical sprays . . ." (1958), op. cit.

P. 251, l. 1: R. Schönmann: "Entomologie und Naturschutz," *Natur und Land*, Vol. xxxviii, No. 5–6 (1952), pp. 61–3.

P. 251, l. 4: W. H. Drury, Jr.: "What about aerial spraying?", *Bulletin of the Massachusetts Audubon Society*, Vol. xlii, No. 6 (1958), pp. 306–8.

P. 251, l. 22: P. F. Springer: "Mosquito control and wildlife," *Wildlife in North Carolina*, Vol. xxii, No. 6 (1958), pp. 14–16.

Concerning insects and insecticides, by Lorus J. Milne and Margery Milne: "The insects gain in their war with man," *The New York Times Magazine*, December 1, 1957, pp. 68–9, 71.

CHAPTER 18 · *Devastated Lands*

P. 252, l. 6: W. E. Howland: "Dams," in *New International Yearbook, 1957* (New York: Funk and Wagnalls; 1958).

P. 257, l. 22: L. D. Leet: *Causes of Catastrophe* (New York: Whittlesey House; 1947).

P. 258, l. 16: Krakatoa Committee: *Report* (London: Royal Society of London; 1888).

P. 258, l. 23: R. D. M. Verbeek: *Krakatoa* (Batavia; 1886).

P. 259, l. 8: Leet: op. cit.

P. 259, l. 12: H. Hansen, editor: *The World Almanac 1959 and Book of Facts* (New York: New York *World-Telegram and Sun;* (1959).

P. 260, l. 2: A. L. Day and E. T. Allen: *The Volcanic Activity and Hot Springs of Lassen Peak* (Washington, D.C.: Carnegie Institution; 1925).

P. 261, l. 6: I. C. Russel: *Volcanoes of North America* (New York: The Macmillan Co.; 1924).

P. 266, l. 13: A. W. Hartman: "Fire as a tool in southern pine," in *Trees: Yearbook of Agriculture, 1949* (Washington, D.C.: U.S. Government Printing Office; 1950), pp. 517–27.

P. 266, l. 16: H. H. Chapman: "Natural areas," *Ecology*, Vol. XXVIII, No. 2 (1947), pp. 193–4.

P. 266, l. 19: C. H. Stoddard: *Essentials of Forestry Practice* (New York: Ronald Press; 1959).

P. 266, l. 25: R. B. Cowles: "Starving the condors?", *California Fish and Game*, Vol. XLIV, No. 2 (1958), pp. 175–81.

P. 268, l. 12: C. E. Randall: "Forest Service, U.S.," in *New International Yearbook* for 1956, 1957, 1958 (New York: Funk and Wagnalls; 1957, 1958, 1959 respectively).

P. 268, l. 28: W. E. Howard, R. L. Fenner, and H. E. Childs, Jr.: "Wildlife survival in brush burns" *Journal of Range Management*, Vol. XII, No. 5 (1959), pp. 232–4.

P. 270, l. 13: J. Adam: "La végétation de la region de la source du Niger," *Annales Géographique* (Paris), Vol. LVI, No. 303 (1947), pp. 192–200.

P. 270, l. 17: P. G. Cross: *Our Friends, the Trees* (New York: E. P. Dutton; 1936).

CHAPTER 19 · *Extinction*

P. 272, l. 9: D. Ripley: "The néné can be saved," *Animal Kingdom*, Vol. LXI, No. 3 (1958), pp. 82–9. Also J. C. Greenway, Jr.: *Extinct and Vanishing Birds of the World* (New York: American Commission for International Wild Life Protection, Special Publication No. 13; 1958).

P. 275, l. 25: G. R. Lamb: "The ivory-billed woodpecker in Cuba," *Pan-American Section, International Committee on Bird Preservation, Research Report* No. 1 (1957), pp. 1–17.

P. 276, l. 24: Editorial, *Science*, new series, Vol. CXXII, No. 3168 (September 16, 1955), p. 509.

P. 276, l. 31: C. van Dresser: "Have the key deer been saved?", *Florida Wildlife*, Vol. XII, No. 3 (1958), pp. 24–5, 38.

P. 277, l. 1: F. Harper: *Extinct and Vanishing Mammals of the Old World* (New York: American Commission for International Wild Life Protection, Special Publication No. 12; 1945).

P. 277, l. 11: J. H. Westermann: *Nature Preservation in the Caribbean* (Utrecht: Foundation for Scientific Research in Surinam and the Netherlands Antilles, Publication No. 9; 1953).

P. 278, l. 7: G. R. Lamb: "On the endangered species of birds in the U.S. Virgin Islands," *Pan-American Section, International Committee on Bird Preservation, Research Report* No. 2 (1957), pp. 1–5.

P. 278, l. 20: Westermann: op. cit., p. 7.

P. 278, l. 25: K. W. Kenyon *et al.*: "Birds and aircraft on Midway Islands November 1956–June 1957 investigations," *U.S. Department of the Interior, Fish and Wildlife Service, Special Scientific Report, Wildlife*, No. 38 (1958), pp. 1–51.

P. 279, l. 12: *Time*, October 26, 1959, p. 74.

P. 279, l. 16: O. L. Austin, Jr.: "The status of Steller's albatross," *Pacific Science*, Vol. III, No. 4 (1949), pp. 283–95.

P. 280, l. 7: J. W. Aldrich: "Conflict of birds and aircraft at Midway," *Audubon Magazine*, Vol. LX, No. 1 (1958), pp. 27–9, 35, 41.

P. 280, l. 23: *Conservation News*, September 15, 1959, p. 9.

P. 280, l. 29: O. L. Austin, Jr.: report in *Sokkojiho* (Japan), Vol. XXI, No. 8 (1954), p. 232 ff., quoted in Greenway: op. cit.

P. 280, l. 35: Greenway: op. cit.

P. 281, l. 13: Greenway: op. cit. Also J. Williams: *Fall of the Sparrow* (New York: Oxford University Press; 1951).

P. 281, l. 20: Editorial, "More sea otters," *Natural History*, Vol. LIX, No. 5 (1950), pp. 236–7.

P. 283, l. 15: R. D. Jones, Jr.: "Present status of the sea otter in Alaska," *Transactions of the 16th North American Wildlife Conference* (1951), pp. 376–83.

P. 283, l. 17: R. S. Griffith: "What is the future of the sea otter?", *Transactions of the 18th North American Wildlife Conference* (1953), pp. 472–80.

P. 285, l. 1: E. S. Huestis, Provincial Fish and Game Commissioner, Edmonton, Alberta, Canada: personal communication, June 2, 1958.

P. 285, l. 17: R. F. Shuman: "Bear depredations on red salmon spawning populations in the Karluk river system, 1947," *Journal of Wildlife Management*, Vol. XIV, No. 1 (1950), pp. 1–9.

P. 25, l. 29: W. K. Clark: "Seasonal food habits of the Kodiak bear," *Transactions of the 22nd North American Wildlife Conference* (1957), pp. 145–51.

P. 287, l. 15: W. E. Scott: *Silent Wings* (Madison: Wisconsin Society for Ornithology).

P. 288, l. 10: L. Keimer: "Animals of Egypt," *Egypt Travel Magazine*, Vol. VIII (1955), pp. 8–13.

P. 289, l. 9: L. M. Talbot: "Marco Polo's unicorn," *Natural History*, Vol. LXVIII, No. 10 (1959), pp. 558–65.

P. 289, l. 18: H. Horwood: "The people who were murdered for fun," *Maclean's Magazine*, October 10, 1959, pp. 27, 36–43.

P. 289, l. 19: Williams: op. cit.

P. 289, l. 29: W. M. Wheeler: "The dry rot of our academic biology," *Science*, new series, Vol. LVII, No. 3 (1923), pp. 61–71.

CHAPTER 20 · *Sanctuary*

P. 291, l. 13: F. Harper: *Extinct and Vanishing Mammals of the Old World* (New York: American Commission for International Wild Life Protection, Special Publication No. 12; 1945).

P. 292, l. 12: L. M. Talbot: "The lions of Gir: wildlife management problems of Asia," *Transactions of the 22nd North American Wildlife Conference* (1957), pp. 570–9.

P. 292, l. 31: D. Allen: *Our Wildlife Legacy* (New York: Funk and Wagnalls; 1954), p. 234.

P. 296, l. 3: Editorial, "A city home for wildlife," *Orange Disc* (Gulf Oil Co.), May–June 1958, pp. 22–5.

P. 296, l. 17: N. M. Dodge: "Running wild," *National Parks Magazine*, Vol. XXV, No. 104 (1951), pp. 10–15.

P. 296, l. 20: T. Eberhard: "Food habits of Pennsylvania house cats," *Journal of Wildlife Management*, Vol. XVIII, No. 2 (1954), pp. 284–6.

P. 296, l. 24: G. C. Toner: "House cat predation on small animals," *Journal of Mammalogy*, Vol. XXXVII, No. 1 (1956), p. 119.

P. 297, l. 12: State of New Jersey vs. Elmer Mayberry, Warren County; 1959.

P. 300, l. 13: W. Vogt: "Unsolved problems concerning wildlife in Mexican national parks," *Transactions of the 10th North American Wildlife Conference* (1945), pp. 355–8.

P. 302, l. 34: A. S. LeSoef: "The wild fauna of Australian cities," *Journal of the Society for the Preservation of the Fauna of the Empire*, new series, Vol. LXI (1950), pp. 33–6.

P. 303, l. 5: V. Cahalane: "Wildlife and the national park land-use concept," *Transactions of the 12th North American Wildlife Conference* (1947), pp. 431–6.

P. 303, l. 12: M. Burton: *Living Fossils* (London: Thames and Hudson; 1954), p. 257.

P. 303, l. 17: E. H. Tong: "World register of Pere David's deer," *Proceedings of the Zoological Society of London*, Vol. CXXIX, No. 3 (1957), pp. 343–9.

P. 303, l. 25: J. Zabinski: "Le bison d'Europe en Pologne," *Bulletin de la Société National d'Acclimatisation de France 1946*, No. 3–4 (1946), pp. 103–6.

P. 303, l. 29: E. Mohr: "Development of the European bison during recent years and present state," *Journal of the Society for th Preservation of the Fauna of the Empire*, new series, Vol. LIX (1949), pp. 29–33.

P. 303, l. 33: J. Zabinski: "Conclusions obtained from twenty years of bison breeding in Poland," *Journal of the Society for the Preservation of the Fauna of the Empire*, new series, Vol. LIX (1949), pp. 11–28.

P. 304, l. 8: Burton: op. cit., p. 222.

P. 304, l. 24: *The New York Times*, March 29, 1959.

P. 305, l. 20: D. Hillenius: [A nature reserve for *Alytes obstetricans* at Epen, Netherlands] *Levende Natuur*, Vol. LIX, No. 7 (1956), pp. 145–9.

P. 306, l. 30: H. D. Thoreau: *Walden* (Boston: Ticknor and Fields; 1854).

Index

i

INDEX

iv

When Canadian-born Lorus J. Milne came to Harvard for his graduate work in biology, he acquired not only his Ph.D. but also his biologist wife, a Radcliffe Ph.D. He is now professor of zoology at the University of New Hampshire, where Margery has also taught.

Lorus and Margery Milne's knowledge of the earth's wildlife heritage has been won by personal observation on field trips covering more than 330,000 miles all over the world, including three expeditions under the flag of the Explorers Club. They are widely known to lecture audiences here and abroad, and to readers of *The Atlantic Monthly, American Scholar, Audubon Magazine, Natural History, Scientific American,* among other periodicals. Their books, numbering a round dozen, have been translated into several languages, and have won distinguished literary awards for the Milnes. Their wildlife photographs, too, have been published in this country and abroad. Television viewers have seen the Milnes more than once in the United States, Canada, and the United Kingdom, on the network program of adventure, *Bold Journey.*

August 1960

A NOTE ON THE TYPE

The text of this book was set on the Linotype in JANSON, a recutting made direct from the type cast from matrices long thought to have been made by Anton Janson, a Dutchman who was a practicing type-founder in Leipzig during the years 1668–1687. However, it has been conclusively demonstrated that these types are actually the work of Nicholas Kis (1650–1702), a Hungarian who learned his trade most probably from the master Dutch type-founder Dirk Voskens.

The type is an excellent example of the influential and sturdy Dutch types that prevailed in England prior to the development by William Caslon of his own incomparable designs, which he evolved from these Dutch faces. The Dutch in their turn had been influenced by Claude Garamond in France. The general tone of the Janson, however, is darker than Garamond and has a sturdiness and substance quite different from its predecessors.

This book was composed, printed, and bound by Kingsport Press, Inc., Kingsport, Tennessee. Paper manufactured by P. H. Glatfelter Co., Spring Grove, Pa.

Typography by Vincent Torre.